The Biology of Soft Shores and Estuaries

Colin Little

OXFORD
UNIVERSITY PRESS

OXFORD
UNIVERSITY PRESS

Great Clarendon Street, Oxford OX2 6DP

Oxford University Press is a department of the University of Oxford.
It furthers the University's objective of excellence in research, scholarship,
and education by publishing worldwide in

Oxford New York

Athens Auckland Bangkok Bogotá Buenos Aires Calcutta
Cape Town Chennai Dar es Salaam Delhi Florence Hong Kong Istanbul
Karachi Kuala Lumpur Madrid Melbourne Mexico City Mumbai
Nairobi Paris São Paulo Singapore Taipei Tokyo Toronto Warsaw

with associated companies in Berlin Ibadan

Oxford is a registered trade mark of Oxford University Press
in the UK and certain other countries

Published in the United States
by Oxford University Press Inc., New York

First published 2000

A catalogue record for this book is available from the British Library

Library of Congress Cataloging in Publication Data
Data applied for

ISBN 0 19 850427 6 (Hbk)
ISBN 0 19 850426 8 (Pbk)

1 3 5 7 9 10 8 6 4 2

Typeset in Baskerville
by J&L Composition Ltd, Filey, North Yorkshire
Printed in Great Britain
on acid-free paper by
T.J. International Ltd

The Biology of Soft Shores and Estuaries

Biology of Habitats
Series editors: M.J. Crawley, C. Little, T.R.E. Southwood, and S. Ulfstrand

The intention is to publish attractive texts giving an integrated overview of the design, physiology, ecology, and behaviour of the organisms in given habitats. Each book will provide information about the habitat and the types of organisms present, on practical aspects of working within the habitats and the sorts of studies which are possible, and include a discussion of biodiversity and conservation needs. The series is intended for naturalists, students studying biological or environmental sciences, those beginning independent research, and biologists embarking on research in a new habitat.

The Biology of Rocky Shores
Colin Little and J.A. Kitching

The Biology of Polar Habitats
G.E. Fogg

The Biology of Ponds and Lakes
Christer Brönmark and Lars-Anders Hasson

The Biology of Streams and Rivers
Paul S. Giller and Björn Malmqvist

The Biology of Mangroves
Peter J. Hogarth

The Biology of Soft Shores and Estuaries
Colin Little

Preface

Marine soft sediments, estuaries, and brackish waters are places of extraordinary biological interest, and home to an immense diversity of plants and animals. Salt marshes and mangrove swamps, for example, are areas with some of the highest rates of primary production in the world. Estuarine flats provide vital feeding grounds for flocks of migrant wading birds and nursery grounds for fish. Coastal lagoons contain relatively few species, but many of these are rarities. Yet such areas, and estuarine mudflats in particular, are often perceived as waste lands, ripe for industrial development or for reclamation. Mangrove swamps still retain the flavour of a comment made in Punch 70 years ago: 'Have you ever been in a mangrove swamp, my dear—Nature at its *most* revolting'. Much of this distaste probably has to do with the properties of the mud that characterizes estuaries and swamps. Not only is it visually unappealing, but for humans it is a glutinous mess, attaching itself rigorously to skin and clothing. It can also be highly dangerous because of its depth, our inability to walk over its surface, and our consequent struggles through it. But its natural inhabitants respond to mud quite differently. For them it represents a three-dimensional habitat offering abundant food, protection from predators, and relative physical constancy. Mud-dwelling organisms exploit the properties of mud instead of complaining about them, themselves remaining clean and keeping the mud particles under control. Because of the importance of soft sediment as a habitat, I have chosen to start this book by discussing the structure of sediments in relation to the lives of the organisms that live within them. Following this, I take up the variety of themes imposed by different kinds of soft sediment, from exposed sandy beaches to vegetation-rich salt marshes and mangrove swamps. I then go on to consider the effects of salinity, and the special characteristics of estuaries. Salinity has in the past been regarded as the major determinant of biological distribution in estuaries, but in my view it is better regarded as 'merely one more brick in the wall'.

The background and rationale for this book have been derived from many years of running lecture and field courses, and from my research interests. I have been extremely fortunate in being able to work in a variety of estuaries and lagoons, and to have found many able and entertaining collaborators. When floundering in a metre of estuarine mud as the tide rises, reliable teamwork and a sense of humour are essential! For help in the field, and in discussing estuarine and lagoonal problems, I thank in particular Richard Barnes, the late

Charlie Boyden, Dick Crawford, Andrew Dorey, Chris Mettam, Don Morrisey, Dave Morritt, Dave Paterson, the late Lynda Smith, Penny Stirling, Graham Underwood, and Gray Williams. For help with literature and sources of illustrations, I thank in addition Frank Round, Simon Thrush, and Marian Yallop. The book has been immeasurably improved by comments on the manuscript from Richard Barnes, Peter Hogarth, Cathy Kennedy, Beth Okamura, Dave Paterson, and Sir Richard Southwood, and I am most grateful to them all, while myself retaining all responsibility for errors and for opinions offered.

C.L.

Bratton
June 1999

Contents

1 Introduction: organisms, sediments, and water movements

During winter, the intertidal mudflats of the Chesapeake Bay, a large estuarine system on the east coast of North America, are crowded with wading birds. Dunlin (*Calidris alpina*), dowitchers (*Limnodromus* spp.), and oystercatchers (*Haematopus* spp.) accumulate in large feeding flocks. Meanwhile, mudflats in the Severn estuary, in southwest England, are visited by up to 2000 curlew (*Numenius arcata*), 2000 redshank (*Tringa totanus*), and 50 000 dunlin. As well as these mobile armies, the richness of the habitats is emphasized by the extensive fringing salt marshes, where cordgrasses (*Spartina* spp.) dominate meadows that are flooded by sea water at high spring tides. These salt marshes are some of the most productive areas in the world, vying with tropical rain forests and coral reefs for the highest rates at which carbon is fixed.

Closer views of the Chesapeake Bay and the Severn estuary reveal that besides the large and obvious organisms such as birds and angiosperms, there is a bewildering diversity of invertebrate animals and microscopic plants. In the Chesapeake, fiddler crabs (*Uca* spp.) roam over the flats while mudsnails (*Ilyanassa* spp.) plough through the surface of the mud in search of algal-rich areas or accumulate in hundreds around the bodies of stranded fish or crabs. In the Severn, the shells of another mudsnail, *Hydrobia ulvae*, may accumulate in enormous densities. These snails are only about 5 mm long, but have been recorded on some mudflats at up to 300 000 individuals in 1 m^2. On the very surface of the mud, unicellular algae, often mainly diatoms but sometimes photosynthetic euglenoids, may form a brown or green film where 20 000 cells are found in each square centimetre.

Careful observation and trapping at high tide shows an influx of predatory fish such as flounders (*Pleuronectes* spp.) and crabs such as *Carcinus* (in the Severn) or *Callinectes* (in the Chesapeake). *Callinectes* is so successful that east-coast American fisheries may catch 50 000 tonnes of them in a year. Observation within the sediments provides further evidence that both the Severn estuary and Chesapeake Bay are extremely rich and productive areas: worms, molluscs, crustaceans, and other invertebrates can be found in profusion.

In contrast to mudflats, open sandy beaches such as those of the Oregon coast in western USA, or those of southern Africa and Australia, show little obvious signs of life to the casual observer. True, there are burrowing amphipods or 'beach fleas', and maybe ghost crabs such as *Ocypode* in tropical areas and near the high tide mark. Lower down on the beaches are more burrowing invertebrates: amphipods, isopods, bivalve molluscs, and polychaete worms. In many areas, olive shells (*Oliva* and *Olivella*) plough through the sand scavenging. On American and African coasts, the burrowing mole crabs, *Emerita*, strain out organic matter from the surf: some of the beaches have dense populations of surf-zone diatoms, leading to high rates of production. But while there may be some scavenging gulls, and occasional flocks of waders feeding at the tide mark, where dead seaweed or 'wrack' has been thrown up by the waves, there are seldom large concentrations of predators to be seen, nor are there any dense concentrations of rooted plants.

Do muddy estuarine shores and open sandy beaches really represent different ecosystems, or are there basic underlying similarities? As we shall see later, the flow of energy may be very different, but the lifestyles of the inhabitants bear many features in common. A major part of this book is devoted to discussing the ways in which sediment-dwelling organisms are adapted to their environment.

Life in soft shores versus life on rocky shores

One way of emphasizing the similarities of life in sandy shores, protected bays, and estuaries, is to contrast it with life on rocky shores. Perhaps for organisms the primary difference between all particulate shores and most rocky shores is one of dimensions. Rocky shores are mostly two-dimensional environments, except where heavy growths of algae form vertical canopies, while particulate shores offer three dimensions (Fig. 1.1). Thus for organisms on rocky shores there is often no escape from predators: many rocky-shore animals are sessile, and their defence can exist only in terms of heavy armament, as for example in barnacle or limpet shells. In sand or mud, however, animals can retreat into the depths when predators appear. Even 'sessile' animals like fan-worms or razor shells can vanish into their burrows many centimetres from the surface at high speed. When the fauna fails to take complete shelter, it seems that many can survive what is called 'partial predation'. This means, for example, that a flatfish may spend much of its time cropping the protruding siphons of sedentary bivalves—but the bivalves do not die; they merely re-grow their siphons. The ability to move around within the sediment also means that it is easier to avoid direct competition with neighbours and it is possible to escape predation by other burrowers. One of the other beneficial effects of sediments is that the finer ones at least usually retain a large amount of water at low tide. Death from desiccation is therefore not such a problem as it is on rocky shores, except high on the shore in coarse sands. At high tide, the sediment acts as a buffer against changes in salinity, temperature, and pH that may occur in the overlying water. Yet another bonus is that because organic materials usually end up as small particles, they accumulate in sediments—so it is often possible to make a living

Fig. 1.1 Comparison between a rocky shore (two-dimensional) and a soft shore (three-dimensional). The rocky shore is based on a moderately exposed situation in northwest Europe. The soft shore is based on the Chesapeake Bay, northeast America. (Partly after Lippson and Lippson, 1997.)

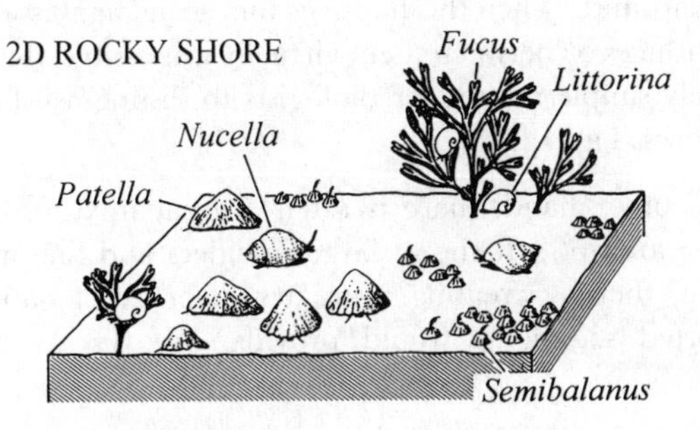

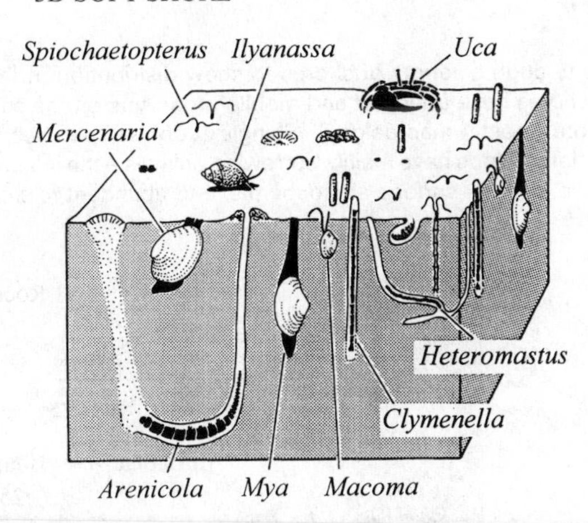

simply by eating the sediment, especially in the finer muds. Very few rock-dwellers can do this, unless the rock is permeated with burrowing algae.

There are, however, also some hazards for organisms living in sediments. First, there is nothing to anchor to, unless the organisms happen to be very small, like microalgae or bacteria. And second, the particles are far from stable so that while a particular patch of shore may be here today, it may be gone tomorrow. This calls for flexibility of lifestyles, and particularly flexibility of feeding behaviour. As we shall see later, many organisms that live in sediments have more than one way of obtaining food. From all these points it follows that the biology of sedimentary shores is in many ways quite different from that of rocky shores.

In spite of these differences, it has been pointed out that there is a continuous gradient of particle sizes on shores, from solid rock through large boulders to shingle, sand, and mud (Raffaelli and Hawkins, 1996). Yet this gradient is not really reflected in biological communities, because intermediate sizes of sediment such as shingle are usually 'intertidal deserts', in which macrobiota are absent. This circumstance arises because shingle shores, on which particles may range from something like 5 mm to 250 mm diameter, are usually only deposited where currents are quite fierce, and where the particles continue to be moved around. When this happens they grind against each other, making life on their surfaces or between them virtually impossible. This 'desert area' makes it a relatively simple matter for biologists to distinguish between rocky shores and soft shores (Fig. 1.2).

On the other hand, there is often a great mixture of sediment types within any one area of soft shore: large boulders and fine muds *do* sometimes occur together, thereby creating a greater diversity of habitats for organisms than well-sorted sediments would provide. We discuss the processes of sorting later.

Fig. 1.2 Sections through a variety of shores to show distribution of fauna and flora. Rock and boulder shores have attached and mobile organisms on the surface—the epifauna and the epiflora (mostly macroalgae). Shingle (cobbles, pebbles, and granules) has little fauna or flora. Sands have mainly burrowing animals—the infauna—while muds (silts and clays) have infauna, surface angiosperms, and abundant microalgae.

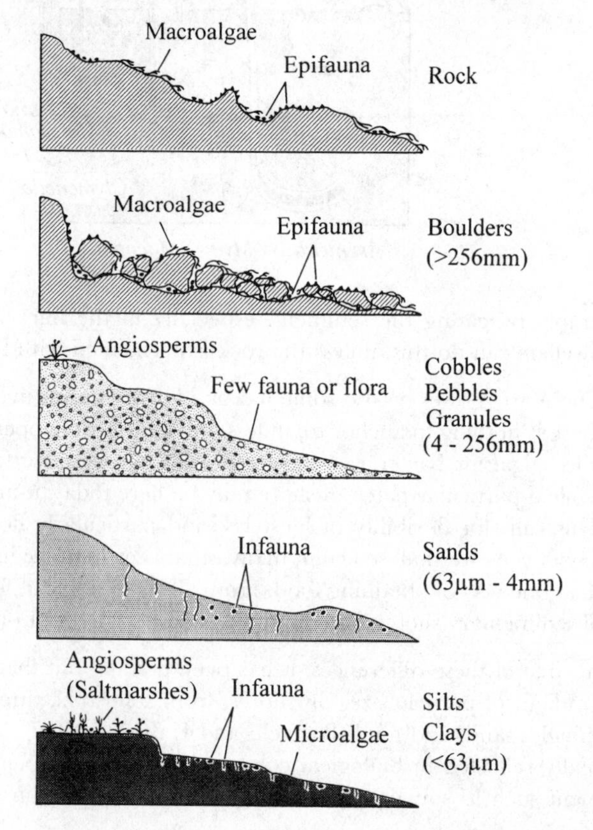

The distribution of sediments

As indicated by the previous discussion, some understanding of the distribution of different types of coastline—from rock to shingle and sand to mud—is essential for an understanding of the distribution and ecology of fauna and flora. A simplistic approach here might suggest that soft sediments occur only where currents are slack, while fierce wave action and strong currents produce a rugged cliff line with coarse sediments. This, however, is to ignore the more immediate question of the *supply* of sediment. Sand and shingle, for instance, may be derived from present-day erosion of cliffs, but enormous volumes also come from deposits formed during glacial and post-glacial times, many of which are found offshore. Thus many coastlines around the world that face heavy wave action—the west coast of America, or the west coasts of some of the Scottish islands, for example—have few rocky cliffs and are lined by continuous sand beaches.

The continued existence of such sand beaches hides the point that they are in dynamic equilibrium: sand is usually on the move *along* the beach, and if its supply is interrupted, or transport systems become more effective, the beach may disappear. In St Kilda, an island in northwest Scotland exposed to extensive wave action, one sand beach vanishes each winter during storms, only to reappear next spring when the gales subside. The movement of sand is demonstrated more commonly, if less forcibly, by its accumulation on the windward side of structures such as groynes that project out from the shore. Only at the top of sand beaches, beyond the reach of normal tides, are rooted plants effective as stabilizing forces.

The situation with muds is also not entirely straightforward. True, mud particles do accumulate where currents are low—in bays, inlets, estuaries, and in sheltered regions generally. But the mud may be derived from many sources: some is brought in from the sea, some comes down rivers, while a great deal that moves around in the turbid waters of estuaries is merely being re-worked from one site to another. Again, the mudflats are far from being static entities: they represent areas of changing balance between erosion and deposition. So although estuaries are often regarded as 'sheltered', mudbanks within them may be eroded and transported elsewhere quite frequently, while salt marshes grow out from the shore only to be cut back when the sediment supply dries up or wave and current climates change. As we will emphasize later, though, plants are very important in stabilizing fine sediments. Organisms from the scale of diatoms up to seagrasses and mangroves can be effective in binding mud particles together.

Assuming there is a local supply of sediment, a combination of waves, tides, and currents is indeed responsible for determining where sediments are deposited. Heavier sediments, like coarse sands and shingle, mostly occur on wave-exposed outer coasts, but where tidal currents in estuaries are strong they may form extensive banks in what appear to be relatively sheltered waters. This is particularly obvious in estuaries with a high tidal range. In the Bay of Fundy in eastern Canada, for example, ridges of sand several kilometres in length lie parallel to the tidal currents at the Bay's outer regions. These sands are so mobile that they support very little animal or plant life. Many British estuaries have beds of sediments containing enough gravel and shell fragments to allow settlement by the mussel *Mytilus edulis*. Commercially exploited mussel beds are thus often a feature of estuaries.

Where the currents and waves are diminished, finer sediments accumulate, so that extensive mudflats occur only where there is some degree of shelter. This shelter may be provided by inlets such as estuaries, and the majority of mudflats occur there. But longshore sandbars also provide protection, so mudflats are certainly not *restricted* to estuaries. As we noted earlier, also, they are seldom entirely static. The turbid colour of estuarine waters demonstrates that much fine sediment is periodically suspended above the bottom. Deposited muds and silts contain a much higher fraction of organic material than the coarser sands, and thus provide a richer food supply for animals. So it is not surprising to see that, overall, mudflats are centres of biological productivity while sandflats are not.

From this very brief introduction, it should be apparent that in order to understand the basic processes that occur on sedimentary shores, whether they occur in estuaries or elsewhere, we need to understand something more about the physical processes that determine the distribution and movement of particles. We therefore end this chapter with a brief discussion of the types of water movement that can resuspend and transport particles.

Water movements

The movements of sediments on exposed sandy beaches are controlled mainly by the forces of wave action, while in sheltered bays and estuaries waves are less important and tidal forces predominate. The influences of waves and tides are not mutually exclusive, but to simplify explanations, we take them in turn.

Waves and the movement of sediment on sandy beaches

Breaking waves on a shore are the most obvious manifestation of water movements in the sea. But do waves really move sediment from place to place? To answer this question we need to consider some basic facts about waves.

Waves in the sea are caused by wind blowing over the sea surface. Winds generate a 'shear stress', and water moves in response to this: the higher the wind speed, the greater the wave height. Waves travel enormous distances without losing much of their energy, so the prevalence of waves arriving at a shore depends to a great extent upon the uninterrupted distance over which the wind has been blowing—the 'fetch'. Thus shores facing, say, into the Pacific from the west coast of America are subjected to high-energy waves, and are said to be very exposed. Shores in the Gulf of Mexico have a much smaller fetch. Within individual inlets such as Chesapeake Bay, fetch will be very small, only low-energy waves arrive, and shores are sheltered.

Over deep water, waves may form a variety of shapes, from sinusoidal, in which each crest is a smooth curve, to trochoidal, in which each has a sharp crest (Brown *et al.*, 1997). This variety of types disguises the point they all have in common: as the wave passes, each 'particle' of water moves in a circular orbit, and the resultant lateral movement of the water is almost nil: the wave moves (and energy is transmitted) but the water does not. As waves pass into shallow water, this situation changes because of the friction between the water particles and the bottom. Where the water depth is less than half the wavelength—the distance between wave crests—the water particles move in an elliptical orbit instead of a circular one. When the waves reach further inshore, and water depth becomes less than one-twentieth of the wavelength, the ellipses at the sediment surface become horizontal lines, and water particles show a to-and-fro movement onshore and offshore (Fig. 1.3). If the force resulting from water flow exceeds the threshold of force required for sediment movement, sand particles will also move to-and-fro in a direction at right angles to the shoreline.

When wave movement directly causes movement of sediment, it results in some degree of sorting: onshore wave velocities tend to be higher than offshore velocities, so coarser (heavier) particles tend to be moved onshore by incoming waves, and not removed in the backwash. Finer sediments, on the other hand, stay in suspension and are moved offshore. Wave action therefore often results

Fig. 1.3 Comparison between deep-water and shallow-water waves. Water 'particles' move in orbits which progressively flatten in shallow water. (After Pethick, 1984, and Brown *et al.*, 1997.)

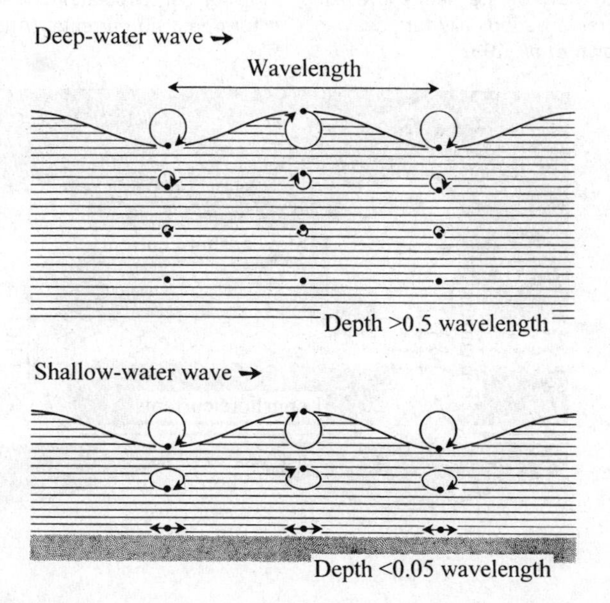

in accumulation of coarser sands at the top of the shore while finer sands accumulate lower down. The consequences for distribution of burrowing organisms can be overwhelming.

A further point about the paths taken by water particles near the shore is that even when they move in ellipses, the ellipses themselves are not quite symmetrical, so that water particles, besides showing a to-and-fro movement, do slowly move towards the shore. If this movement continued, we could imagine a nonsensical situation where water piled up on the shoreline. Instead, water moves sideways, along the shore, and then at intervals the excess water moves offshore again in narrow bands. These are called rip currents, and they may have quite high velocities. A kind of circulation cell is thus formed, and sediment moves within this cell (Fig. 1.4).

Waves contribute to movement of sand along the shore in other ways as well. Most waves do not strike the shore at right angles, but come in obliquely, although usually at angles of less than 10°. This oblique movement creates a longshore current which may reach 1 m/s on gently sloping beaches. On steep shores, a strong wave swash will move obliquely up the shore, while the backwash usually runs down directly normal to the shore, so sediment travels along the shore in a saw-tooth path. Overall, longshore transport of sand and shingle can be substantial: net movement of material southwards past one point on the southeast coast of the USA may be as much as 500 000 m^3/year.

Fig. 1.4 Plan view of the near-shore zone, showing cell circulation in which slow longshore currents periodically turn seawards to form fast rip currents. (After Pethick, 1984, and Brown *et al.*, 1997.)

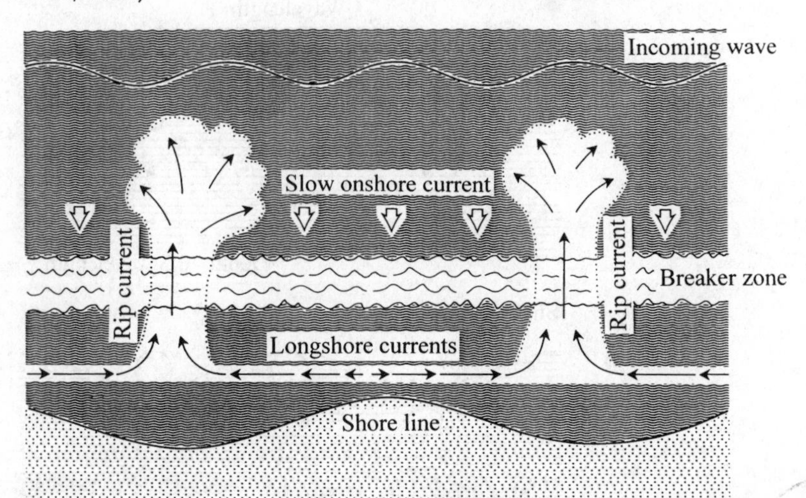

Tides, tidal currents, and the movement of sediment on mudflats

Because the movement of the tides is slower than that of wind-driven waves, it is far less obvious. But in enclosed areas such as estuaries, tides are responsible for violent currents and consequently for the movement and sorting of enormous quantities of sediment.

Tides are caused by the gravitational pull of the moon and the sun, which act on the world's oceans (Pethick, 1984; Brown *et al.*, 1997). As the earth rotates, a tidal wave moves around it. The moon, being nearer, provides most of the energy input driving the tides, so the frequency at which tides rise and fall at any one point on the earth relates primarily to the position of the point in relation to the moon. The net result is that in a period of 24 h and 51 min, a fixed point experiences two high tides and two low tides. In other words, tides occur roughly twice each day, but are later each day by a period of 51 min (Fig. 1.5).

The height that the sea rises and falls over a tidal cycle—the tidal range—varies enormously. Partly this range depends upon the relative positions of earth, moon, and sun. When all three bodies are lined up, tidal range is largest—the so-called spring tides. When the forces exerted by the sun and the moon are operating in opposition, tidal range is smallest—the neap tides. But tidal range

Fig. 1.5 Comparison of a microtidal regime (Bergen) with a macrotidal regime (Cherbourg). The tidal levels shown are mean high water springs (MHWS), mean high water neaps (MHWN), mean low water neaps (MLWN) and mean low water springs (MLWS). From information in the Admiralty tide tables.

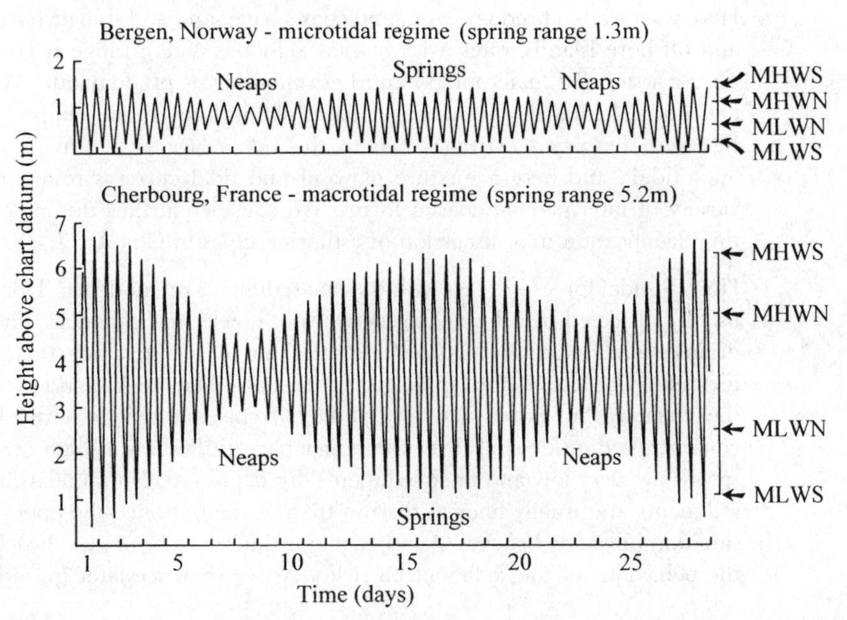

is also dependent upon geography: as the tide wave moves around the earth, it reacts to frictional forces in much the same way as wind-driven waves reaching a shore. The tides in open oceans seldom have a range of more than 2.5 m, but in coastal inlets the tidal amplitude can be greatly magnified. In funnel-shaped estuaries such as the Bay of Fundy in North America, and the Severn in southwest Britain, the frequency of the tidal wave coincides closely with the natural resonance period of the inlets, and tidal range reaches as much as 15 m.

The important point here in terms of sediment movement is that as tidal range increases, higher volumes of water move into and out of areas such as estuaries, so that tidal currents increase. In turn, the potential for sediment movement increases. To complicate matters, tidal currents will change not only in velocity but in direction over the tidal cycle, so sediment may move in a very tortuous path which is difficult to predict. But the general result is that, although estuaries and inlets may be relatively calm in terms of wave action, they are often subject to strong current action, and the currents have a strong influence on sediment deposition and sorting. For instance, flood-tide currents produced as the tide rises are usually stronger than those when tide falls (the ebb-tide currents). So sediment that is moved into the estuary on the flood may be left behind on the ebb: an estuary may act as a sediment trap. Consequently, when considering the supply of sediment to an estuary, it is not sufficient to measure input from rivers because there may be substantial input from the sea.

The relative importance of wind-driven waves and of tidal currents to a great extent determines coastal landforms and sediment distribution. In particular, as tidal range is a measure of the strength of tidal currents, it is often a good descriptor of coastal processes. Tidal range can usefully be classified into three band widths (Fig. 1.5). Where range is less than 2 m, it is termed 'microtidal'. Here wave action predominates, and coasts have sand and shingle barriers, spits and offshore islands, often with coastal lagoons. Where range is greater than 4 m, it is termed 'macrotidal', and here tidal forces predominate. Wide mudflats and salt marshes occur, particularly within large funnel-shaped estuaries. Between these two extremes, where the tide range is 2–4 m, it is termed 'mesotidal', and here a mixture of wind and tidal currents results in a wide variety of intermediate coastal forms. We consider further the importance of this classification in a discussion of estuarine types in Chapter 7.

How do tidal forces interact with the fine sediments on mudflats? Tidal flats are usually dissected by a branching network of creeks and channels. These act as drains, taking water off as the tide falls, but it is also along them that the incoming tide rises before spilling over the flats. The creeks and lower parts of the flats may thus experience strong tidal currents, and tend to have a high sand content. As the tide spills over the upper part of the flats at high tide, current speeds are very low and finer sediments are deposited. On mudflats, therefore, sediments are usually finer at the top than at the bottom—the opposite to the situation on sandy beaches. Another major difference from sand beaches lies in the behaviour of the sediment particles. As we shall see later (p. 59), silt and mud particles clump together and do not behave as individuals like sand particles. The result is that they are hard to erode, and high-shore mudflats in particular are relatively stable.

The overall balance between tidal forces and the forces of wave action greatly influences the sedimentary regime in which soft-shore organisms live. Should this balance change, or should there be a change in sediment supply, the beach may erode, accrete, or change in sediment composition. We go on to consider how organisms and sediments interact in Chapter 2.

2 The world of particles: a variety of habitats

The conditions within different types of sediment vary enormously, and as the lives of organisms living on soft shores are intimately bound up with the characteristics of the sediments, we begin by considering sediments as habitats.

Sediments as places to live

Most of the animals that live on soft shores spend their time below the surface of the sediment—these are the infauna. Plants, on the other hand, must at least keep part of their structure in the light for photosynthesis. Nevertheless, for plants as well as for animals, conditions within the sediments are crucial because they affect such factors as the supply of water, oxygen, and nutrients, and the stability of the system. Many of these factors are in turn determined by the types and sizes of the particles involved in the sediment, and the degree to which the particles are sorted. This is because size and sorting determine how 'open' the sedimentary environment is, and therefore how much water flows through it, bringing with it fresh supplies of oxygen.

A comparison of two sediments—a coarse sand and a much finer one—illustrates these points. In both these sediments, conditions change with depth, but the changes are slight in the coarse sand and dramatic in the finer sand. Perhaps the simplest crude measure of change is the redox potential (Eh), a measurement which reflects the balance between oxidation and reduction processes (Fig. 2.1). Near the surface, the redox potential is positive, showing that oxygen is present and that oxidation is the primary chemical process. Lower down the potential decreases and then becomes negative, showing a reducing environment. The oxidized surface sediments are yellow or brown while the reduced ones are black. In between is a grey layer in which the redox potential decreases rapidly—this layer is called the redox potential discontinuity (RPD). The depth of the RPD reflects how rapidly oxygenating water passes through the sediment. In coarse sand it may be metres down, and change in redox potential may take place over a considerable vertical distance. In muds and fine sands the RPD may be only a millimetre down and is often extremely abrupt.

Fig. 2.1 The redox profile in the Nivå Bay, Denmark, showing also the vertical distribution of five ciliates. Yellow sand is well oxygenated. The grey sand is equivalent to the RPD layer (the redox potential discontinuity), and below this is a layer of decomposing *Zostera* (a seagrass). Some of the ciliates live in black sand below this. (After Fenchel, 1969.)

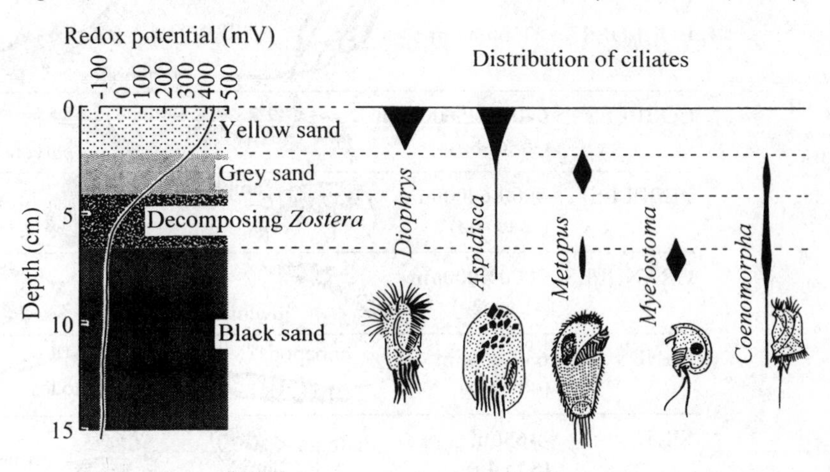

In the oxic zone, above the RPD, more oxygen arrives or is generated than the organisms—including those ranging in size down to bacteria—consume. Animals in this zone have 'normal' aerobic metabolic processes like those of surface dwelling species. In the reducing zone, below the RPD, organisms face quite different conditions. There is little if any free oxygen (see p. 17 for further discussion), and there are high levels of hydrogen sulphide and ammonia, both of which are toxic. Animals in the reducing zone therefore have to employ anaerobic processes, or provide their own oxygen supply, as discussed later.

Because particle size determines some of the basic properties of sedimentary environments, we begin by considering how particle sizes are sorted and how variations in particle size affect the biota.

Particle size, sorting, and consequences for the biota

One of the problems in studying sediments is that the particles that make up sediments have an enormous size range. Coarse wave-washed cobbles may be over 20 cm in diameter, while clay minerals may measure less than 1 μm—a range of more than five orders of magnitude. One way to express this range is to use a logarithmic scale, and the most widely used in the past has been the Wentworth scale. Here particle size is expressed as the negative log to the base 2 of diameter in millimetres, and this function is called phi (ø). Note that finer sediments have large (positive) values of phi while coarse sediments have negative values. However, many biologists and sedimentologists now find it simpler to quote actual sizes in microns.

For convenience, various size fractions have been given names, so that words like 'silt' and 'pebble' have now acquired a defined meaning (Fig. 2.2). The

Fig. 2.2 The size range of sediments, compared with organisms of equivalent sizes. Particles are categorized on the Wentworth scale with ø units in brackets.

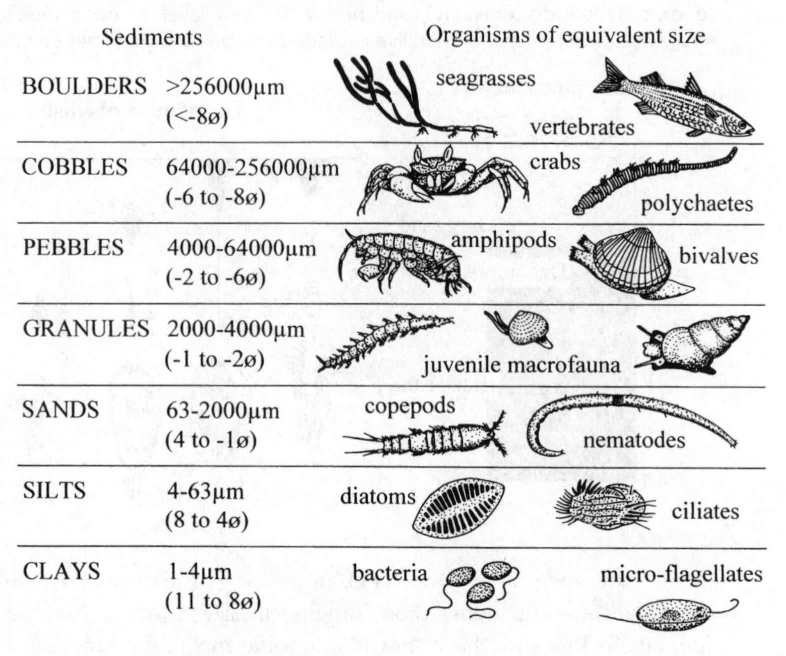

word 'mud' is taken to include all particles less than 63 µm (i.e. silts and clays combined). The distinction between muds and sands is particularly important when considering sediment composition. The clay particles in muds have electrochemically charged sites that attract ions and allow hydrogen bonding—the particles therefore bind together, and muds have what are called 'cohesive' properties (p. 59). Sands, in contrast, if they do not contain many clay particles, do not bind together.

When the proportions by weight of the various size fractions have been determined, they can be plotted either as histograms or as a curve showing cumulative percentage (Fig. 2.3). It is then easy to compare different sediments by comparing their different cumulative percentage plots. Alternatively, as a shorthand, the median particle diameters (ø50) of the sediments can be compared. Further statistics that define the degree of sorting can also be extracted, by reading off ø at specific levels such as 25 and 75% or 16 and 84% (see, e.g. Gray, 1981).

Very often, it is impossible to analyse all the size fractions of a sediment completely, but a good idea of its properties can still be obtained by estimating just the proportions of finer particles. Thus if the proportion of particles less than 250 µm is high, drainage will be poor and oxygen concentrations will be low. If the proportion of particles less than 63 µm (the silt-clay fraction) is high, the organic fraction is likely to be high, and the sediment will also usually show 'cohesive' properties, as we have already mentioned.

Fig. 2.3 Particle-size composition of two sediments from a brackish-water lagoon, shown as cumulative percentage frequencies. Although the two sediments have values of ø50 showing they are predominantly coarse, both have more than 15% of sand finer than ø = 1.0 (i.e. 500 μm). (After Barnes *et al.* in Jefferies and Davy, 1979.)

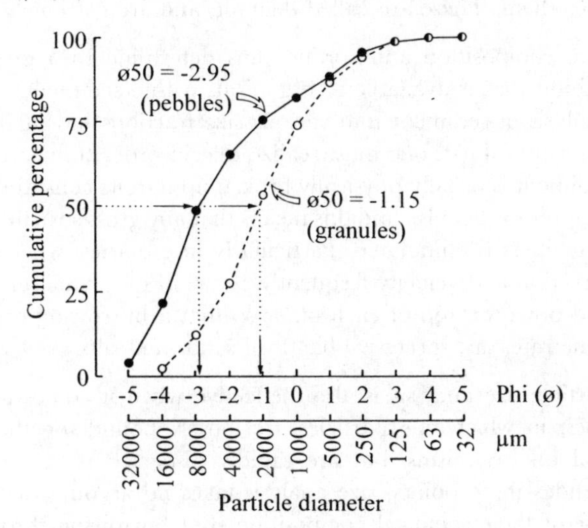

Much of the importance of the finer fractions is due to the fact that the higher the proportion they represent, the lower is the amount of space between the particles. In an ideal sediment of spherical particles all of the same size, the space between them—known as the porosity—would be 26% of the total whatever the size of the spheres. Smaller particles added to such a sediment would fill in the gaps between the particles and would therefore reduce the porosity.

The importance of porosity, in turn, is mainly that it determines the permeability—the rate at which water can drain through the sediment. In general, if the proportion of fine grains is high, permeability is low. For organisms living below the surface of a sediment, this means there will be a low oxygen content. The organisms have either to adjust to this, or have to bring in oxygenated water directly.

Porosity is also important in determining the density of the sediment bed. Sediments with a high porosity—and therefore a high water content—have a low density. These sediments may behave like fluids, and they can be relatively unstable unless bound together by biological activity (p. 25). When porosity is low and density high, the sediment is compacted and generally hard—such sediments behave as solids, are difficult to erode, and often have restricted biological activity. Under certain conditions, a sandy sediment can behave either as a solid or as a fluid in response to an outside force. An example of this is a quicksand. Under normal conditions, the bed seems solid, but if disturbed (e.g. by vibration) the particles lose contact with each other and become suspended in the sediment's pore water. The bed then behaves as a fluid. Such sediments are called thixotropic. Humans find these extremely inconvenient, but burrowing animals can tunnel into them easily. In contrast, some sands harden under pressure, seen as a whitening when someone walks over them. These are called dilatant, and are extremely hard to burrow in.

Size composition and sorting thus determine to a great extent the kind of habitat that is available to organisms within sediments. But, unfortunately, an analysis of sediment into various size fractions has many failings as a habitat descriptor. First, one must realize that in order to measure these fractions, the sediment is usually physically broken up into its constituents (see the end of the chapter for details), and this means that any gross structuring such as layering or patchiness is eliminated. Particularly in estuaries, where currents may change direction and velocity frequently, layers of quite different particle sizes may be laid down on top of each other so that a burrowing organism would have to penetrate, say, successive bands of sands and silts.

Particle-size analysis is also ineffective when it comes to the measurement of flocs, in which fine particles are loosely bound together. Flocs may be very real for organisms, but are rapidly dispersed if the sediment is broken up. Besides these points, size analysis takes no account of the shape of particles, nor of their chemical composition. But burrowing through a bed of angular particles may be very different from slithering between rounded sand grains. A bed of fine inorganic clays may have the same size composition as a bed of organic debris—but the possibilities for a food supply in the two beds are quite different!

Finally, particle-size analysis obscures the point that between the particles there may not be a 'void' filled with water. Many sediments are now known to contain some kind of organic matrix between the particles, varying from a granular composition (Watling, 1991) to more uniform arrangements of mucopolysaccharides (Paterson, 1989). These matrix substances may bind the sediment particles together and completely alter its characteristics, as we shall see later.

Thus, although the determination of particle-size fractions in a sediment is a useful first step, it is unlikely that this type of technique will tell us very much about what conditions are really like in the sediment. In this respect, various recent studies give a view of sediments more appropriate to the ways in which infaunal organisms must experience them. Techniques using thin sections of sediments (Watling, 1991) and others in which the sediment is frozen and then examined by scanning electron microscopy (Paterson, 2000) show the distribution of the grains, and demonstrate that while some particles are clean, others are covered by an organic film. Between particles may be the organic matrix mentioned above. These techniques may even show the distribution of bacteria, but fail to produce explanations for such distributions. Bacteria are usually attached to the particles yet the surfaces of particles are very far from covered by bacteria. In 1 g of mud in which the grains have an average diameter of 10–50 μm, the particles have a total surface area of between 3 and 8 m^2. There might be as many as 10^9 bacteria in the same weight, but these would occupy only 1.4% of the surface area available. Why is so much surface area uninhabited?

We turn first to sediment chemistry, which can at least tell us something about conditions *between* the particles. There are many aspects to interstitial chemistry, but for the moment we leave aside consideration of salinity, as we shall discuss this in detail in later chapters, and concentrate upon the importance of oxygen and of situations where it is in short supply.

Oxygen and hydrogen sulphide

The supply of oxygen to the infauna is without doubt the most important chemical influence on the biology of sediments. As we have seen, the amount of oxygen present is reflected by the sediment's colour, and the black layer below the RPD is usually described as anoxic. This is not quite true, as we shall see. What it really means is that oxygen is often at such low concentrations that sulphate-reducing bacteria such as *Desulphovibrio* can exist and multiply. These and others use sulphate as a hydrogen acceptor, and in so doing produce hydrogen sulphide (H_2S). H_2S itself is a colourless (but very strong-smelling!) gas, but it reacts with oxides of iron to form iron sulphides. It is these which are black and colour the layer below the RPD. On mudflats where the RPD is near the surface, filamentous bacteria such as *Beggiatoa* oxidize the sulphide to elemental sulphur to fuel their own metabolic processes. White or yellow patches can then sometimes be seen over the anoxic sediments.

Because H_2S is formed by biological action, its rate of production varies with factors that affect metabolic rates. In summer, when temperatures are high, the rate of infaunal respiration rises, so oxygen is consumed faster, while bacteria also show greater activity. Oxygen levels therefore decline, in spite of increased oxygen production by photosynthetic microalgae, and the RPD moves up in the sediment. In winter, the reverse happens and the RPD moves downwards.

If the zone below the RPD is totally anoxic, can organisms live there? Most macrobenthos (i.e. organisms larger than about 500 μm) can only live there if they have burrows or tubes to provide an oxygen supply. But smaller organisms, the meiobenthos (100 μm–500 μm) and microbenthos (less than 100 μm) are often abundant (Fig. 2.1). The meiobenthos, in particular, have species that are characteristic of habitats with low oxygen and high sulphides, and these have been called the 'thiobios' (Fenchel and Finlay 1995). There has been much discussion about these animals, ranging from suggestions that they form a primitive and archaic ecosystem, to conclusions that they form a continuum with those that require oxygen, the 'oxybios'. This discussion has been enlightened by recent observations with oxygen and sulphide microelectrodes, which have shown that the region below the RPD is far from uniform (Meyers *et al.*, 1987). Oxygen levels are generally thought to be negligible, except where oxygen is imported by burrowers, and in any case, imported oxygen is rapidly used up. Some species of meiofauna probably are limited to the zones around burrows, where there is at least sometimes an oxygen supply. But many meiofauna can tolerate anoxia for long periods, and have definable distributions which appear to be mainly based on the amount of sulphide present. Among the small flatworms, for example, 60% of the *Parahaploposthia* populations are found in sulphide-rich sediment, while *Turbanella* is found where sulphide is very low. Claims by some authors that all meiofauna are limited to areas where oxygen is supplied from the burrows of macrofauna are therefore certainly not true. Indeed, several nematode species have now been described which contain symbiotic sulphur bacteria, and these probably can exist without a supply of free oxygen. Other species do not seem to be governed totally by either oxygen or sulphide, so there are almost certainly other chemical factors involved.

According to Meyers *et al.* (1987), the microhabitats defined in terms of oxygen and sulphide levels are very small—no more than two to three times the size of the animals—so the patchiness of habitats must form a very fine-grained mosaic. It has been shown that chemical gradients exist on an even finer scale, controlling bacterial distribution. As so often in biological studies, it is easy to over-simplify when studying the conditions in which organisms live.

Organic material

The supply of organic matter to a sediment may depend upon local sources of production such as diatom mats on the sediment surface, or upon imports from other sites of production. The latter may include a wide variety such as salt marshes or mangrove swamps, algae torn from rocky shores, river-borne material derived from the land, or plankton in overlying water. Whatever its source, however, this material has been derived by breakdown of animal and plant material, and is called detritus. Detritus has been formally defined as the organic carbon lost from any trophic level, but excluding losses by predation (Fenchel and Jørgensen, 1977). There is much discussion (see p. 79) concerning the way in which this detrital material is used by deposit feeders, but at this stage it is important to note that it *is* a valuable source of food to the infauna. It is also immensely important in influencing the conditions within sediments.

This is because, while inorganic particles are slowly broken down by the physical processes of erosion, organic particles are also continually being broken down by a variety of biological processes. A classical example is shown by breakdown of the leaves of cord-grass, *Spartina*. In the early stages of decay, leaves lose soluble organic substances by leaching. A dead leaf landing on a sediment surface therefore provides material that stimulates rapid bacterial growth. Mechanical action of waves, and trituration by animals that feed on dead leaves then reduce the leaves to fragments, causing the exposed surface area of leaf cells to increase. The increase in surface area allows colonization by more bacteria, protists and fungi, which use the organic matter as a substrate.

What is happening to the nutritive content of the leaves during these decay processes? This question may be approached by looking at the proportions of carbohydrate and protein in the leaf fragments while they break down. As the fragments decrease in size, carbohydrate content decreases, but perhaps surprisingly the protein content *increases*. By the time the leaf fragments are in the state of fine particles, protein may reach 30% of their dry weight. The protein increase occurs because although leaf proteins decline, bacterial and fungal populations build up, producing their own protein. A crude measure of this process can be gained by measuring the C/N ratio. Initially C/N is high, perhaps 20, because carbohydrates contain little nitrogen and there is little nitrogen-rich protein. As decay proceeds, the ratio falls to something like 10, indicating high levels of (microbial) protein (Mann, 1988).

C/N ratios have thus often been used to suggest how much useful food there is in sediments. Unfortunately, while this ratio is often a good guide, it can also be misleading. Carbon, for instance, is often present in the form of coal, which distorts the picture. Similarly, it is probably more informative to measure protein or amino acid concentration than total nitrogen. It may also be essential to measure bacterial biomass, as some hypotheses suggest that deposit feeding macrofauna gain most of their energy by eating bacteria and not by eating 'dead' detritus. We shall return to this question on p. 80.

As detritus degrades into smaller and smaller particles, it also has a physical effect on sediment, reducing porosity. Thus sediments of high organic content tend to have low permeability and hence low oxygen concentrations. In particular, the fine-grained sediments tend to accumulate more organic matter than coarse ones. Muds and silts therefore often become anoxic rapidly with depth because their low porosity reduces oxygen supply and because bacterial activity on the organic particles increases oxygen consumption. As we have seen, the RPD in fine muds and even sands may be only a millimetre below the surface.

Particles of detritus are usually referred to as 'particulate organic matter' (POM). But besides the particulate fraction, there is also a substantial portion of organic matter in solution—dissolved organic matter (DOM). Many organisms, both benthic and pelagic, are now known to be able to take up DOM and utilize it, and the concentration of DOM in interstitial water may be crucial for some species. Even classic deposit feeders like the polychaete *Clymenella* may be able to satisfy a good part of their metabolic needs with DOM, and we will discuss its importance in Chapters 3, 4, 5 and 10.

How organisms affect sediments

In the above section we have seen that organisms form an integral part of the sediment habitat. It should not be surprising, therefore, to find that organisms can themselves affect the structure and chemistry of the sediment, and thus alter their own micro-environments, and those of others. Some organisms act to bind sediment together, others loosen sediment structure and are called 'bioturbators', while many have complex effects that may act in both directions. In an effort to simplify the situation, we will consider disturbance and stabilization separately, but it is important to remember this separation is rather artificial.

Bioturbation or disturbance

One of the widespread ways in which macrofauna affect the sediment is by forming burrows. For example, some of the polychaete worms can construct deep burrows and irrigate them with surface water, thus effectively raising themselves above the RPD. The parchment worm, *Chaetopterus variopedatus*, named from its tough membranous tube, is found on both sides of the north Atlantic. It has specially developed paddles that pull water down from the surface, past the head and out again at the tail end of a U-shaped burrow. Oxygenated water thus moves as much as 20 or 30 cm below the sediment surface.

Perhaps more important are animals that besides providing for themselves, alter conditions near their burrows so that smaller animals can exist in oxic habitats. Good examples of this are shown by the various species of lugworm. In Europe, *Arenicola marina* creates a roughly U-shaped burrow in sand, through which it draws oxygenated water by pumping movements of the body. In this case, water comes in from the tail end and passes out over the head. The increased oxygen in the burrow lining, compared with the anoxic sediment further away, allows a variety of small polychaetes to survive, as well as some of the lesser-known worms such as gnathostomulids (Reise and Ax, 1979).

Lugworms were initially thought to be unselective deposit feeders, eating sediment that fell from the surface into a 'head-shaft', and then defecating at the other end of the burrow. However, in 1975 Hylleberg showed that the faecal casts of *Abarenicola pacifica*, found on the west coast of the USA, actually contained *more* organic matter than the sand which the worm was thought to be eating. He suggested that in the 'feeding pocket', a cavity below the head-shaft, conditions were such that bacteria, flagellates, ciliates, and other microbes could flourish, and that *Abarenicola* combined the eating of these microbes with the selection of fine sediments to satisfy its metabolic requirements. In other words, *Abarenicola* benefited by altering the conditions in the feeding pocket (by increasing oxygen concentration, and thus promoting an inward diffusion of nutrients) and thus could be said to be 'gardening' (Fig. 2.4). To garden is, after all, to control conditions so that desired plants, and not the ones that would occur naturally, shall grow. This is exactly what Hylleberg suggested *Abarenicola* is doing, except that it cultures microbes, and as we shall see later, many animals on soft shores may be thought of as carrying out some form of gardening. Note that in the case of *Abarenicola*, conditions are altered in the feeding pocket *and* in the faeces, where organic material is concentrated, stimulating yet further growth of microbes.

The effects of burrowing deposit feeders are not limited to changes in organic content and oxygen supply. Many burrowers also affect sediments by re-working

Fig. 2.4 The lugworm *Abarenicola* in its burrow, 'gardening' microbes by aerating sand in its feeding pocket, which also derives nutrients from the surrounding anoxic sand. (After Hylleberg, 1975.)

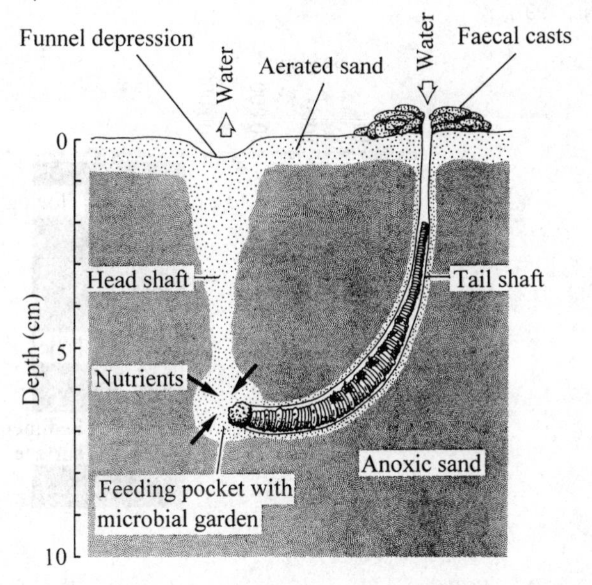

them. The bamboo worm, *Clymenella torquata*, is common on the east coast of the USA, and makes vertical tubes in sandy mud. Here it lives head downwards, and it eats sediment at depth, then defecates at the surface. As it prefers to eat fine particles, these accumulate in a layer at the surface, where conditions are thus substantially altered. This is far from being the whole story, however, as studies on the related worms *Praxillella* spp. have shown (Levin *et al.*, 1997). *Praxillella* feeds head-down, but while moving in its burrow and raking sediment upwards, it also disturbs sediment near the surface, which falls as a 'rain' of particles to the feeding cavity some 10–13 cm below (Fig. 2.5). As the worms are very common at some sites, especially offshore from North Carolina, they may be responsible for transferring large quantities of material—especially fresh POM—into the depths of the sediment. The supply of organic material to regions inhabited by deposit feeders with no access to the surface is obviously very important, and it has been proposed that the activities of *Praxillella* control much of the community structure and dynamics in the relatively deep-water sediments in which it lives.

The rates at which deposit feeders can re-work sediment are extraordinary, and even suspension feeders may confer gross changes on wide areas. The deposit feeding bivalve *Macoma tenta* produces layers of faecal pellets on the surface of the sediment, which may reach 5–10 mm thick if undisturbed. On the other hand, the rate of re-working can vary enormously within a species in different areas: *Macoma* may re-work anything from 5×10^{-5} to over 600 cm^3/day—a range of nine orders of magnitude (Hall, 1994). Another bivalve, *Nucula annulata*, re-works sediment up to five times in 1 day. Populations of the suspension

Fig. 2.5 The maldanid worm *Praxillella* 'hoeing' sediment from the surface, some of which falls to the bottom of the tube. The graph shows densities of fine beads (105–149 μm), 1.5 days after they were added to the surface. Some of these are found below 10 cm. (After Levin *et al.*, 1997.)

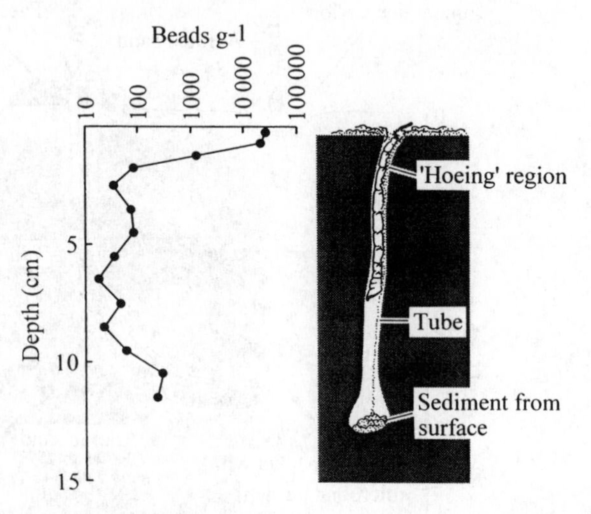

feeding bivalve *Cerastoderma (Cardium) edule* in the Dutch Waddensee produce 100 000 tonnes of material each year in the form of faeces and pseudofaeces—material rejected but bound together with mucus. This re-processing of deposited or suspended material has a major effect on sediment structure. Primarily, it increases the sediment's water content, sometimes nearly doubling it. Thus a community of deposit feeding bivalves, including *Macoma tenta*, increased water content from 40% to 70% (Rhoads and Young, 1970). The resulting sediment was much less stable, and more thixotropic, allowing easier burrowing.

In the tropics and south-temperate zone, some of the major bioturbators are crabs. Sheltered muddy-sand beaches in New Zealand are crowded with the burrows of *Helice crassa*. These may reach down as far as 50 cm, and the crabs can often be seen bringing fresh sediment to the surface. Exposed beaches in the tropics may support ghost crabs, *Ocypode* spp., which burrow in dry sand; while more sheltered beaches may have armies of soldier crabs, *Mictyris* spp., which do not form permanent burrows but corkscrew themselves into damp sand at high tide. In Tasmania, soldier crabs burrow down to 15 cm, disturbing the sediment and breaking up the normal gradient of oxygenated and reduced sand to such an extent that the whole community structure of the meiofauna is altered (Warwick *et al.*, 1990). Fiddler crabs, *Uca* spp., occur on sheltered beaches and salt marshes from Australia and southern Africa to east- and west-coast USA, and most of their activities are centred on burrows, which may be more than 1 m deep. These species and the many others found on sandy shores must have very significant effects on sediment properties. As many of them are deposit feeders, ingesting large quantities of the sand itself, they must have direct effects on distribution and sorting of surface sediment as well.

In mangrove swamps, crabs probably have even greater effects. Most crabs burrow below the surface of the mud, or inhabit burrows made by others. The extent of this burrow system is so great that below the mud surface there is a maze of crab-runs, and the effect of this maze on water flow and oxygenation is enormous (Ridd, 1996). As the tidal water flows through a swamp in Queensland, Australia, water flow through the burrows reaches between 1000 and 10 000 m^3 in each square kilometre over one tidal cycle. This flow must enhance exchange of nutrients between the substratum and overlying water, because the sediments are very fine-grained and have low porosity.

The consequences of this exchange have yet to be elucidated in mangroves, but in European mudflats the polychaete worm *Nereis diversicolor* can increase flux rates dramatically (Hansen and Kristensen, 1997). Addition of *Nereis* to cores of mud caused an increase in the sediment's rate of uptake of oxygen; and while part of this increased uptake was due to the worm's metabolism, most was due to increased microbial activity. At the same time, rates of efflux of carbon dioxide and ammonia increased. These changes occur because *Nereis* forms burrows which it irrigates, increasing exchange.

Animals that live within the sediments are not the only ones that affect conditions below the surface. Large predators that live on the surface or visit from the water above may create pits by digging for prey. On a small scale, fish such as flounders (*Pleuronectes* spp.) take bites of a few square centimetres, while at the other extreme, gray whales (*Eschrichtius robustus*) may excavate areas of up to 6 m^2 (Hall, 1994). The pits or disturbed areas cause patchiness in the physical environment, alter the flow regime over the surface, and may be a major cause of faunal heterogeneity in sediments. Further patchiness may be caused by a quite different mechanism—the accumulation of drifting dead algae on the sediment surface (Thrush, 1986). Decaying algae create an anoxic region beneath, raising the RPD and altering conditions for the infauna. Causes of patchiness are discussed in more detail in Chapter 6.

Effects of disturbance on faunal distribution

In some cases, disturbance of the sediment by one species has been shown to affect other species directly. For example, when the lugworm *Arenicola* is added to azoic areas of sediment, and re-colonization is studied, the amphipod *Corophium volutator* reappears only where worm densities are low (Hall, 1994). This is probably because re-working of the sediment stimulates *Corophium* to emigrate to more stable areas. Another good example occurs in the Baltic, where the burrowing amphipod *Pontoporeia* kills so many juveniles of the bivalve *Macoma balthica* that the two species scarcely live together as adults (Olafsson *et al.*, 1994). *Pontoporeia* physically disrupts the sediment by its burrowing activities and newly settled *Macoma* are killed.

There may also be positive effects of disturbance—for instance disturbance oxygenates the sediment and allows the smaller infauna to penetrate deeper than they could otherwise. But in general, the effects of disturbance are negative, and it has been suggested that on a wider scale deposit feeders as a whole may decrease the sediment stability so much that they make life impossible for suspension feeders. This negative effect is called amensalism, in contrast to commensalism—a relationship in which one species benefits from another without harming it. Here, however, it is proposed that one whole feeding group has an effect on another, so the idea is known as the 'trophic group amensalism' hypothesis (Rhoads and Young, 1970). This has been favoured until recently because it would neatly explain the situations in which deposit feeders and suspension feeders show mutually exclusive distributions. However, there are now many objections to this hypothesis (Snelgrove and Butman, 1994), including the point that an inverse relationship in distribution is far from universal. Rhoads and Young erected their hypothesis after studying Buzzards Bay, Massachusetts, where one of the common suspension feeders is the bivalve *Mercenaria mercenaria*. Using experimental transfers, they showed that when this species was placed near the bottom in a muddy habitat, it grew more slowly than in its normal sandy area. Presumably its suspension feeding mechanisms were clogged by the constant re-working of sediment by deposit feeders. But in fact it is now known that bioturbation can be just as rapid in sand, where suspension feeders are common (Hall, 1994). In any case, early ideas of splitting animals into discrete groups of suspension feeders and deposit feeders are now thought to be too simple. As we shall see, many species can switch between these two modes. Nevertheless, deposit feeders and suspension feeders do respond differently to conditions produced by flowing water, and in many cases these different responses result in mutual exclusion. Wildish and Kristmanson (1997) have therefore postulated a 'trophic group mutual exclusion' theory.

One major problem in discussing effects of organisms on sediments is that, as Snelgrove and Butman (1994) have pointed out, 'sediment instability' has rarely been clearly defined. Although sediment re-working may lead to 'unstable' bottoms, it would be better to define conditions in hydrodynamic terms. In the future, collaboration between biologists and hydrodynamicists may be the way forward.

Stabilization

Disturbance of sediment, or bioturbation, is widespread. But organisms living in the sediment may also have the opposite effect, that of stabilization. Perhaps surprisingly, some of the most effective stabilizers are micro-organisms. Both bacteria and diatoms secrete material known as mucilage. This material consists of carbohydrate polymers which are usually referred to as 'extracellular polymeric substances' (EPS). As a single bacterium of *Azotobacter* sp. has been calculated to produce enough EPS in one day to coat 500 particles of 0.4 μm diameter, and as EPS is sticky, the effect on particle adhesion can be substantial. It follows that resistance to erosion by water flow will be increased.

The secretion of EPS is also a major characteristic of mobile or 'epipelic' diatoms (Fig. 2.6). In order to move, these microscopic algae exude EPS, and as in muds and silts diatom populations can rise to more than 10^4 cells/cm^2, the EPS may fill the entire void space between the particles, as well as forming a mucilaginous layer on the sediment surface. In muddy sediments, epipelic diatoms can affect sediment erosion, as shown by an experiment in the Severn estuary, England (Underwood and Paterson, 1993). Here one site was treated with a biocide to remove diatoms, while a nearby control site was untreated and had prominent diatom growth. When fresh sediment was deposited by the tide, it was retained for more than a week on the control site, but was rapidly eroded away where the diatoms had been killed by the biocide.

In sandy sediments, epipelic diatoms are not so common, and are joined by 'epipsammic' forms, which attach to grains and move very little. These diatoms may stick grains together by adhering to more than one with their EPS. But the major stabilizing force in coarse sediments is produced by filamentous cyanobacteria ('blue–green algae') which form a physical network of threads (Fig. 2.6). Forms like *Microcoleus* and *Oscillatoria* are common in sands from the Netherlands

Fig. 2.6 Profiles of the surface of sand in Texel, the Netherlands (top) and of mud in the Severn estuary, England (bottom). In sand, a bubble-mat of cyanobacteria and diatoms such as *Navicula* and *Achnanthes* oxygenates the top 4 mm, and primary production occurs in the top 1.5 mm. Extracellular polymeric substances (EPS) and algal filaments bind the grains together. In mud, diatoms such as *Nitzschia* are abundant in the top 1 mm and their EPS binds together the mud particles. Only the top 1 mm is oxygenated. (After Yallop *et al.*, 1994).

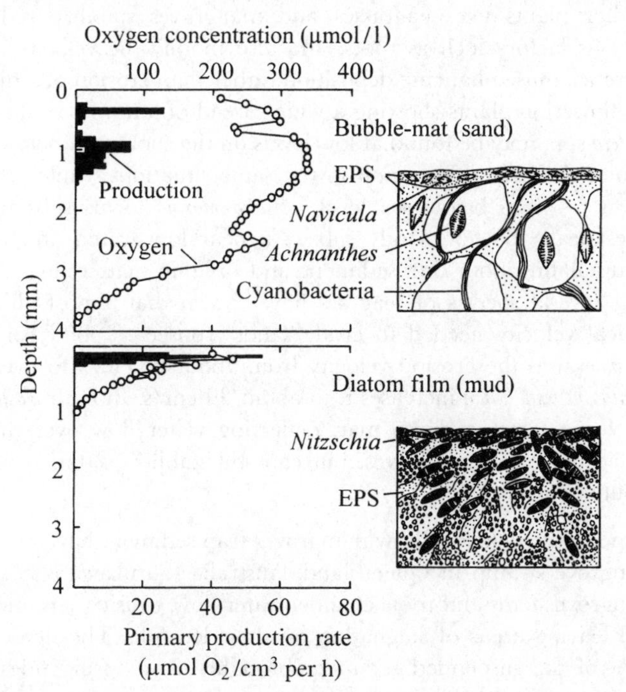

coast to the beaches of Australasia, where they form 'microbial mats'. These may be 1–2 mm deep and strongly resist erosion. Because the algal threads are sticky with EPS, they trap sand grains and form a tough matrix which also smoothes the sediment surface: this smoothing reduces the stress 'felt' by the bed for a particular flow, so the bed is more stable. Filamentous fungi can also bind grains together, with similar stabilizing effects (Meadows *et al.*, 1994).

Cyanobacterial mats affect the sediment below them in other ways. At times they produce so much oxygen that their mucilage forms 'bubble-mats'. But primary production is limited to the top few millimetres, and the sand is often anoxic below this (Fig. 2.6). The restriction of primary production to the surface layers is due to lack of light lower down, and indeed the organisms themselves are responsible for absorbing much of the light.

Examples of muddy sediments with epipelic diatoms and sandy sediments with cyanobacteria are shown in Fig. 2.6. Oxygen is produced down to 1 mm in mud by diatoms, but down to 4 mm below the cyanobacterial mat. In both situations, resistance to erosion is increased compared with clean sediments, but the mechanisms by which this occurs are still not clear. Attempts at relating erodibility of sediments containing mixed natural populations of micro-organisms to simple factors such as various fractions of EPS, or to biomass of algae, have not produced clear-cut correlations (Yallop *et al.*, 1994). But experiments with single-species cultures have shown that sediment erodibility decreases as diatom films grow. In this case, erosion rate is strongly (negatively) correlated with bulk carbohydrate and chlorophyll concentrations (Sutherland *et al.*, 1998).

Higher plants like seagrasses and mangroves stabilize sediments by more obvious factors such as roots, and also modify the velocity and directions of currents, thus enhancing deposition rather than erosion of sediment. Seagrasses are flowering plants showing a wide spread of latitude. In the temperate zone, *Zostera* spp. may be found at low levels on the shore and below tide level. In the tropics *Thalassia* is common in the same situations, while other species reach down to 80 m below sea level. *Zostera marina* forms substantial beds in the Chesapeake Bay on sandy substrata near low water, and is responsible for accumulating both fine sediment and organic matter (Fig. 2.7). Experiments with several species of seagrass have shown that almost all can increase the critical velocity needed to erode sands (Fonseca, 1989). For instance, *Zostera marina* raises the erosion velocity from about 25 cm/s to over 31 cm/s, while *Thalassia testudinum* increases it to about 39 cm/s. In both *Zostera* and *Thalassia*, the leaves form a dense mat, deflecting water flow over the canopy. Other species have too few leaves, but can still stabilize sediment with their underground rhizomes.

Some of the details of how mangroves trap sediment have been described for a mangrove swamp in Queensland, Australia (Furukawa *et al.*, 1997). Here the mangrove stems and roots channel water flow, causing jets and eddies of water, and leaving areas of stagnation on their lee side. The dense mangrove traps 80% of the suspended sediment brought in on spring tides, resulting in an

Fig. 2.7 The change in particle-size composition from bare sand (coarse stipple) to a bed of the seagrass *Zostera* (fine stipple). Finer particles in the *Zostera* bed are due to a mixture of causes: decreased current velocity, addition of organic fragments, and altered infauna. Bars show range. (After Orth in Coull, 1977.)

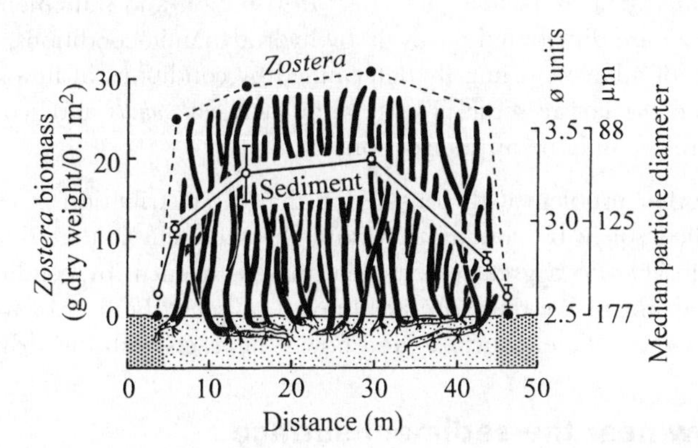

estimated rise in level of the substratum of 1 mm/year. Much of this settlement is due to the formation of flocs, which form because current speeds are low—usually less than 0.2 m/s. Experimental studies on mangroves in New Zealand have shown that the density of aerial roots is strongly correlated with the rate of sediment deposition—and in a swamp near Auckland, this deposition has allowed the mangroves to extend 200 m in a period of 50 years (Young and Harvey, 1996).

In contrast to plants, most animals in sediments create disturbance rather than stability. But tube-building worms in some cases may have positive effects. Polychaetes such as *Pseudopolydora* certainly facilitate recruitment by other taxa, probably because their tubes produce small-scale alterations in the local environment. Whether these alterations are direct effects on binding sediments, or whether the tubes act by affecting water flow is unclear. The tubes of another polychaete, *Lanice conchilega*, act like the steel reinforcing rods in concrete, and increase the rigidity of the sand (Jones and Jago, 1993).

Conditions in water above the sediment

If infaunal species were restricted to narrow ranges of, say, particle size or organic content, it would be reasonable to assume a primary relationship with sedimentary characteristics. As we have discussed above, early studies suggested this was so, and that while suspension feeders were primarily found in coarse-grained sediments like sands, deposit feeders were found only in fine-grained muds. This kind of observation led to the 'trophic group amensalism' hypothesis. More recently people have realized that many species are distributed over a very wide range of sediment types (Snelgrove and Butman, 1994). The suspension feeding bivalve, *Mercenaria mercenaria*, for example, is generally regarded as being more successful in sand than in mud—but in fact *Mercenaria* is common in seagrass beds, where the sediment is usually very fine. If the distribution of adults is not directly related to sediment characteristics, how can we explain it? One suggestion is that adult distribution is determined to a great extent by factors that affect distribution and settlement of larvae. If the larvae are distributed passively by hydrodynamic conditions, then the distribution of adults is primarily determined by conditions of flow above the surface when the larvae settle. The relation between *adults* and sediment type might therefore only be an indirect one.

Another problem with considering that the distribution of the infauna is controlled solely by conditions within the sediment is that these conditions are themselves to a great extent determined, in turn, by conditions in the water above them. We therefore need now to discuss the flow of water in the region near the sediment surface, and how it interacts with the sediment itself.

Water flow near the sediment surface

When water flows over a sediment bed, the rate of flow close to the surface is never the same at all heights above it: friction slows down the flow nearer to the bed, and in fact in a very thin layer next to the bed the flow is actually zero. Above this, the water can be imagined as a series of layers, each moving faster than the one below, and therefore 'shearing' over it. There is therefore a shearing force between each layer, and the bottom layer exerts a shearing force on the sediment—the so-called boundary or bed shear stress. When this exceeds a critical value, the sediment will start to move.

Shear stress is a force, and because it is not very easy to imagine how this relates to water velocity, it is often converted to another term, shear velocity, which is expressed in direct velocity units. The boundary shear velocity is an abstract term—it cannot be measured directly. Nevertheless, it has the advantage that it can be calculated from the velocity gradient above the sediment surface. As with shear stress, there is a critical value of shear velocity above which the sediment will start to move.

The pattern of velocity with height above a bed is shown in Fig. 2.8. The thin layer next to the bed, where velocity is low but shear is high, is called the 'viscous sub-layer'. Above this is a 'logarithmic layer', in which velocity increases, but the rate of increase decreases logarithmically with height. The viscous sub-layer and the logarithmic layer together constitute the 'boundary layer', which extends to a height at which the velocity reaches 99% of the main-stream flow. Above this the flow is unaffected by the bed.

The sediments on the bed generally reflect the characteristics of the boundary layer above. Thus if the region regularly experiences high shear velocities, sediments are coarse; while if shear velocities are rarely high, sediments are fine. However, it is important to note that this will be so only in terms of long-term averages (Snelgrove and Butman, 1994). When current regimes change, it may take a long time before sediments are back in equilibrium, and if the supply

Fig. 2.8 The velocity profile over a bed of sand, recorded in a laboratory flume, showing the boundary layer. Water can be thought of as a series of layers, each shearing over the one below. The extent of the viscous sublayer is shown by the dashed line. The boundary layer extends to a height at which velocity reaches approximately 99% of the main-stream velocity. (After Nowell *et al.* in Lopez *et al.*, 1989.)

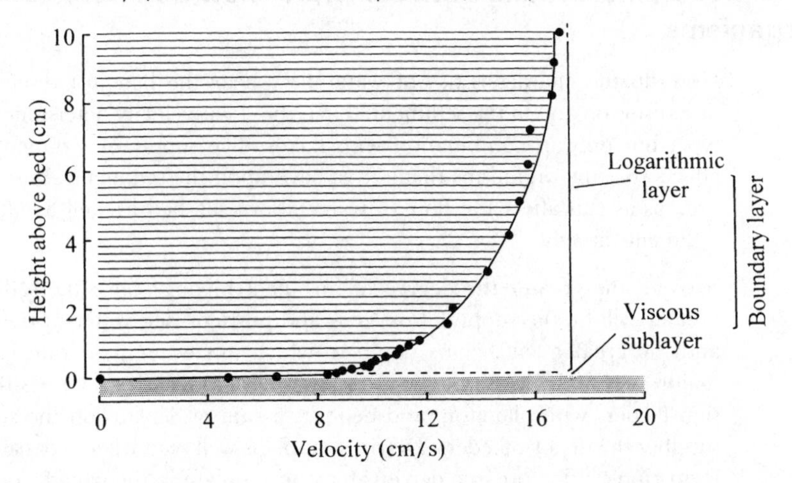

of sediments is restricted, or has been influenced by past conditions which no longer exist, the sediments may *not* reflect the hydrodynamic regime. For instance, fine sediments have accumulated off Nova Scotia, Canada, over geological time, and are still there today *despite* high water velocities near the bed. In addition, intertidal sediments are often influenced by the shear stress caused by breaking waves as well as that caused by currents, so conditions may change rapidly over the tidal cycle.

One further point about the interaction between near-bed velocity and sediment must be emphasized here, and this concerns the roughness of the bed. If the bed is made up of fine sediments, the viscous sub-layer will probably remain intact. In this case, hydrodynamicists speak of 'smooth turbulent flow'. But if grain size is larger, the sub-layer breaks down and turbulent eddies reach right down to the bed: we have 'rough turbulent flow'. This brings more potential for sediment movement, so the same main-stream velocity above a fine sand and a coarse one has more tendency to move the coarse sand. This somewhat counter-intuitive point obviously has important consequences for the infauna.

In fact, fine sediments are less likely to be eroded than coarse ones for two other reasons. First, they are often bound together by microbial mats (p. 25). Second, those with a high proportion of very fine particles (less than 2 μm) form 'flocs' or 'aggregates' which stick so firmly together that they form what are called 'cohesive' sediments. Once these have been deposited, a surface veneer of particles may be lifted off into the water at low shear stress, but the bulk of the sediment requires a very high shear stress to erode it. When bulk erosion does occur, the sediment does not lift off as individual particles but as relatively large lumps. In contrast, coarser sediments are usually 'non-cohesive', and erode as individual particles. We discuss the importance of cohesiveness in more detail in Chapter 4.

Reciprocal interactions between water flow and benthic organisms

How do the characteristics of water flow near the bed just discussed affect organisms on and in the sediment? First, the shear velocity determines sediment type, but only in combination with factors like supply of sediment, and the effects that the organisms themselves have upon the sediment. As we shall see, organisms can affect the flow of water above the bed as well as affecting the sediment directly.

Second, the greater the degree of turbulent mixing above the sediment, the greater will be the supply of oxygen- and nutrient-rich water to the sediment, and the greater will be the depth to which this water penetrates. The other major factor which is brought in the water is food. Water supplies the suspension feeders with plankton, and deposits a rain of detritus on the surface that supplies the deposit feeders. Detritus supply is well seen where coarse sediments form ripples: fine organic detritus may accumulate in the troughs between the ripples, and these troughs often contain accumulations of deposit feeding amphipods (Snelgrove and Butman, 1994).

Flow in the boundary layer will also be critical for distribution and settlement of larvae. There is now growing evidence that some larvae are deposited passively like particles, as mentioned above. But some larvae are known to be able to escape from the viscous sub-layer if conditions in the sediment are unsuitable, and these species may have more specific distributions with regard to sediment composition. Once larvae have settled, of course, there may be many processes of selection that operate to determine survival before the adult population is formed—predation, competition, differential tolerance of physical conditions, food supply, and so on. We discuss some of these factors later.

So far it sounds as if the water flow above the sediment affects organisms without any reciprocal effects of organisms on flow, but this is very far from the truth. As burrowers such as the lugworm *Arenicola* produce pits at the feeding end of their burrows, and mounds where they defecate, they influence flow considerably, with the result that fine organic material accumulates in troughs. The influence of animals on flow may, however, have a much longer chain of influence. Another polychaete, the sand-mason *Lanice conchilega*, builds sandy tubes. As described on p. 27, the lower parts of these tubes directly stabilize the sediment. The upper parts project above the surface of the sediment. At some densities, these tubes may *indirectly* help to stabilize the sediment because they promote turbulence that brings more nutrients to the sediment, and thus promotes the growth of microbes that bind sediment particles together (Eckman, 1985).

Larger organisms such as seagrasses and mangroves that project further from the bed have correspondingly larger effects on flow, as we have discussed above.

In these cases it is often impossible to separate out the effects on stability of the sediment *per se* from the erosive effects of the flow of water over the surface.

Resuspension of sediment

Finally, it is appropriate to discuss one other major interaction between boundary-layer flow over sediments and the organisms that live within them: the matter of sediment resuspension. Unlike rocky substrata, which other than in the long term stay in one place, mud and sand substrata are controlled by the dynamics of deposition and erosion. When shear velocity exceeds a critical value, sediment is eroded, and when it falls below a critical value, sediment is deposited. These two critical values, however, are usually different: particularly for fine sediments, shear velocity for deposition is very much lower than that for erosion. In other words, fine sediments are only deposited when current flow is very slow indeed, but they can resist high flows once in position.

When sediments *are* moved back into the water column, the process is called resuspension. Resuspension is caused by a combination of tidal currents, wave action, bioturbation, and human activities such as dredging and trawling, when a mixture of inorganic sediment, organic particles, bacteria, diatoms, and so forth, moves into the boundary layer. Organisms can influence the degree of resuspension in many ways. Perhaps the simplest situation is where isolated shells or tubes act to increase the roughness of the bed, causing local turbulence—eddies reach right down to the bed, causing scour and sediment suspension.

Depending upon the size fraction involved, sediment may stay in suspension for a matter of minutes to hours (<63 μm) or may be re-deposited in seconds to minutes (>63 μm). On Georges Bank (on the east coast of North America), the critical shear velocity for erosion is exceeded for 62% of the time, so tidal resuspension is a consistent feature (Grant *et al.*, 1997). Such resuspension is thought to provide mobilization of food for suspension feeders—and Georges Bank supports the largest offshore scallop fishery in the world.

The degree of resuspension—and the usefulness of resuspended material to suspension feeders—is greatly influenced by the particle-size composition at the sediment surface. Recent observations suggest that in many areas this is composed of a thin veneer of very fine particles—a kind of 'fluff' layer which has a small standing stock, but a high organic content and enormous surface area for microbial production. In areas where deposit feeders are active, fine material may be mixed with accumulations of faecal pellets. These alter the surface properties greatly: they are larger than individual particles, so they increase roughness and therefore aid resuspension. Pellets produced by the small snail *Hydrobia ulvae* in a tidal creek in the Wadden Sea, Europe, are resuspended to such an extent that in water taken 1 m above the bottom, they reach concentrations of 18 000/l (Edelvang and Austen, 1997).

Suspended particles

Particularly in estuaries that show a high tidal range, where there is usually a rich supply of fine particles, the boundary between deposited and suspended states may be hard to determine. At spring tides (those of high amplitude), fine particles are brought into suspension, so many estuaries have regions where suspended sediment concentration is high—the so-called 'turbidity maximum'. When springs change to neap tides (those of low amplitude), current velocities fall, and the fine sediments are deposited again, forming layers with concentrations as high as 10 g/l. The mixture that results is called fluid mud. Fluid mud may travel considerable distances with the tide. As it tends to become anoxic quickly, its passage over the bottom may have far-reaching consequences for the infauna. When mud suspensions remain static for long periods, they continue to consolidate, reaching concentrations of more than 250 g/l, and may accumulate to thicknesses of metres over regions several kilometres long (Kirby, 1988).

Much of the suspended particle load in the sea, however, is far more widely scattered, and is either of biological origin or acts as a centre for microbial activity. The importance of the activities of bacteria in sea water has only recently been realized, but it is now thought that an essential feature of open-water food webs is the recycling of DOM by a microbial food chain: bacteria, heterotrophic flagellates, ciliates, and so on (Mann, 1988). The centre of this activity is associated with suspended particles.

Some larger suspended particles are formed from aggregates of dead biological material: faecal pellets, organic debris, mucus secretions, diatom frustules, gelatinous remains of pelagic invertebrates. These particles may reach 20 mm or more in diameter and are easily visible when illuminated properly: they form what is called 'marine snow' (Alldredge and Silver, 1988). Aggregated particles are found at all depths, and contain bacteria at densities 2–5 orders of magnitude greater than in surrounding sea water.

Suspended particles have a central position in marine food webs (Fig. 2.9). Classical theories of ocean biology concentrated upon the phytoplankton–zooplankton–carnivore links, and ignored detritus-based processes. Now, however, we know that a large pool of DOM is created by macrofauna and phytoplankton, and this fuels a 'microbial loop': bacteria attached to suspended particles utilize the DOM, and are then in turn eaten by microflagellates which provide food for ciliates (Mann, 1988). The 'loop' provides an extra food supply for zooplankton, and returns DOM to the main food chain. Suspended material does sink slowly to the ocean floor, and the rain of detritus is extremely important in providing a food source for the benthos, especially in the deep sea.

In inshore waters, aggregates also form in suspension, often from inorganic particles that are bound together by electromagnetic forces. The Tamar estuary,

Fig. 2.9 The position of suspended particles in marine food webs, supporting the 'microbial loop'. This loop and the classical food chain are shown as heavy arrows. Thin arrowed lines show dissolved organic matter (DOM). Dashed lines show particulate organic matter (POM).

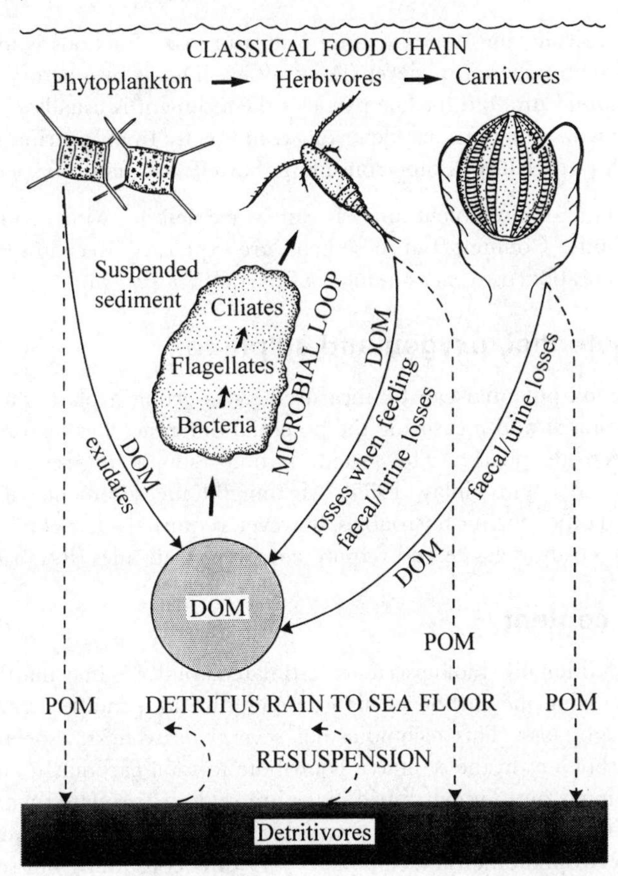

England, for example, forms particles that are more than 500 μm in diameter when current speed is low, but these break up into particles about a tenth this size when the tide ebbs (McManus and Elliott, 1989). The particles have an enormous surface area—1 g may have an area as high as 24 m^2. They have attachment sites for metals, so they may be extremely important when considering dispersion and movement of pollutants. As these particles may move from the sediment surface into the water column and back, in each tidal cycle, their future study will be critical for the understanding of many aspects of sedimentary biology.

Techniques for analysing sediments

Particle-size analysis

Classically, the standard way to measure size fractions is to pass the sediment through a series of sieves (Dyer, 1979). This is satisfactory for particles larger than 63 μm, but for fine particles the sediment is usually re-dispersed in water, and size fractions are determined indirectly by measuring the rate of settling. This procedure is time-consuming but effective and inexpensive.

Automated sediment analysis can be carried out with various modifications of Coulter Counters, but these items are expensive. Recently, laser technology has been applied to measurement of particles *in situ* (McManus and Elliott, 1989).

Redox potential, oxygen and sulphide

Redox potential can be measured by inserting a platinum electrode into the sediment and measuring the potential difference between this and a reference electrode nearby. The system is quite simple to use and relatively robust (Fenchel and Finlay, 1995). Meaningful measurements of oxygen, sulphide, and other chemical variables, however, require the use of microelectrodes which are costly, delicate and require complex electronics (Revsbech *et al.*, 1983).

Organic content

Traditionally, biologists have estimated total organic matter in sediments by oxidizing the carbon at high temperature in a muffle furnace and measuring weight loss. This technique has several drawbacks, especially where coal or carbonates in the sediment contribute to total carbon. In any case, as we have pointed out, faunal distributions are seldom correlated with total POM. It is more realistic to measure a variable like total protein (Mayer *et al.*, 1986). This spectrophotometric technique is very time-consuming but sensitive.

Sediment stability

Measurements of water flow, sediment roughness, and other characteristics that affect erosion of sediments normally necessitate sophisticated equipment. In recent years, much work has been carried out in the laboratory in flumes, where hydrodynamic conditions can be controlled. In addition, the development of portable flumes that can be placed *in situ* on the sediment surface have begun to give real, rather than estimated values (Black and Paterson, 1997).

There have, however, been some developments of small portable devices, and the most successful of these is the 'cohesive strength meter' (Paterson, 1989). This instrument fires a jet of water at the surface of the sediment and then records the amount of suspended sediment produced in a small chamber. Calibration in terms of bed shear stress has been established, and the device has been used on both cohesive and non-cohesive sediments (Yallop *et al.*, 1994).

3 The coarse extreme: life on sandy beaches

Sand and shingle shores are high-energy environments. Waves pound the shores, stirring up and eroding particles, making settled life difficult for plants and animals. The particles are, in any case, non-cohesive, so unless they are held together by the actions of organisms, the sediment is unstable. There are often few signs of life to be seen on the surface, where the particles are most likely to be disturbed: most fauna is infauna. Those organisms that can cope with mobile sediments can exploit the plentiful supplies of oxygen and nutrients available in such habitats. Overall, there is a high diversity of types of organisms living within the sediment, but not a high biomass.

The physical features of such shores determine, to a great extent, the species that live there, and we need to examine a few points about the physical nature of the sandy-shore environment before proceeding.

Some physical features of sandy shores

Sandy beaches stretch between two extremes: where wave action is fierce, the beaches usually have a very gentle slope, and waves generate a wide surf zone offshore, as they reach shallow water. On the coast of Oregon, western USA, this surf zone may reach 300 m in width, and within this zone the waves dissipate much of their energy. Hence such beaches are called 'dissipative'. Sands here are usually fine (less than 200 μm), and there is a rich intertidal fauna burrowing into them. The tide range is usually large, and as the tide rises and falls it pumps water through the sands, renewing oxygen.

At the other extreme, where wave action is relatively mild, beaches tend to have steeper slopes which reflect the waves rather than dissipating their energy in a surf zone. Hence these are known as 'reflective' beaches. Sediment particles here tend to be coarser, and the beach may even be shingle—composed of pebbles (4–64 mm) and cobbles (64–256 mm). The slope of the beach in these cases may rise to a striking 25°. Where the beach is steep, tides are not very effective at irrigating the interstitial spaces below the sand surface, but the incoming swash of each successive wave flushes water through the sand, and even subtidally the pressure changes associated with wave crests and troughs pump water through the interstitial pores. Because the particles are larger than on dissipative beaches, the sediments tend to dry out faster at low tide.

Between these two extremes, the dissipative and the reflective, there is a whole range of intermediates. In fact, most beaches have both dissipative and reflective elements, at different tidal heights. As beaches are also dynamic forms influenced by changing wave climates, beach profiles may change radically with seasons. Thus reflective portions may be eroded by winter storms, when material is moved offshore and the beach is flattened. When wave conditions subside, reflective areas are built up again. In the long run, reflective beaches may be cut back to form flatter profiles.

It follows from this variability even on one beach that it is not always easy to assess the overall degree of exposure to wave action. In practice a number of points can be used to estimate exposure. These include measurement of particle diameter, wave climate, width of the surf zone, and characteristics of the redox potential discontinuity (RPD). Such factors can then be combined into an exposure scale (Brown and McLachlan, 1990).

In conditions of increasing shelter, beaches accumulate more fine particles and higher organic content, and beach profiles flatten out again to form the well-known intertidal flats, which are considered in Chapter 4.

The distribution of sandy-shore organisms

Exposed sandy beaches have a restricted number of common species, but many of these are found worldwide (Fig. 3.1). The burrowing surf-clams, *Donax*, for example, are found near low water in tropical and temperate regions, although not in very cold waters. In New Zealand, *Donax* is absent and the commonest clams are *Paphies* spp. (= *Amphidesma*). Some of the highest concentrations of *Donax*, which is a suspension feeder, are found where there are persistent patches of diatoms, such as *Asterionella* and *Anaulus* in the surf zone. Within the tropics and subtropics, the surf zone may also have dense populations of another suspension feeder, the mole crab *Emerita*. This genus is common on both the east and west coasts of the USA and in Africa, and a related genus, *Hippa*, is also found in Australia. Besides suspension feeders, the surf zone may contain abundant scavengers. The best known is the snail *Bullia* which is found from South Africa through Asia to Australia.

In the middle of the shore, exposed sandy beaches are dominated by bivalves, amphipod and isopod crustaceans, and polychaetes. Some of these have cosmopolitan genera—among the amphipods, *Urothoë* is worldwide and among the isopods so is *Eurydice*. Of the polychaetes, *Nephtys* and *Glycera* occur from the tropics to both north and south temperate zones.

On the upper shore, crustaceans are often the only abundant animals. Amphipod crustaceans often burrow under the drift line and have numerous genera—*Talitrus* in Europe, *Talorchestia* in Australasia, and *Megalorchestia* on the Pacific

Fig. 3.1 The occurrence of some sandy-shore organisms. Black dots show areas where persistent patches of surf diatoms are found.

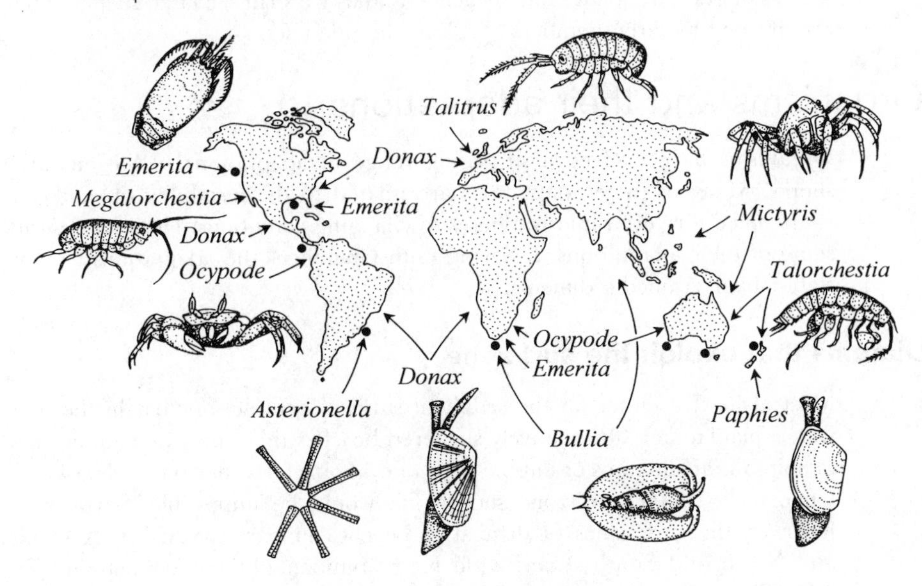

coast of the USA, for instance (Fig. 3.1). In warmer zones they may be joined or replaced by isopods of the genus *Tylos*, closely related to woodlice, and in the tropics and subtropics the upper shore may be dominated by crabs. Large burrowing forms such as *Ocypode*, the ghost crab, are found throughout the tropics.

As exposure decreases, sandy shores show an increasing diversity of species. In particular, deposit feeding polychaetes become dominant, and the genus *Arenicola*, the lugworms, is cosmopolitan. A similar genus, *Abarenicola*, is found on the west coast of the USA and in New Zealand. Deposit feeding shrimps such as *Callianassa* also form burrows in muddy sand, and are found worldwide. Many other invertebrates, such as burrowing sipunculids and further groups of polychaetes and bivalves are abundant. Genera of bivalves such as *Tellina* take over from the surf-clam *Donax*. In east-coast USA, the large suspension feeding bivalve *Mercenaria* is common where conditions are fairly calm, and in many parts of the world the cockles, *Cardium* and *Cerastoderma*, may form enormous populations. Besides these, echinoderms make an appearance. *Echinocardium*, the heart urchin, is a cosmopolitan genus, while sand dollars such as *Dendraster* and *Mellita* are common on the west and east coasts of the USA respectively. Other genera of urchins are widespread in the Indo-Pacific.

Among the predators on sheltered sandy shores are snails somewhat like the scavenger *Bullia*: genera such as *Polinices* and *Natica* can bore through the shells of bivalve prey and extract the flesh, and are common worldwide. Other predators come in a wide variety—on the east coast of the USA, tiger beetles, *Cicindela*, are vicious predators of amphipods, for instance, while bird and fish predators are abundant worldwide. Wading birds such as plovers, and many species of gulls, frequent sandy beaches, and as we shall see later, sheltered flats are invaded by large numbers of fish as the tide rises.

Organisms and their adaptations

Organisms have evolved a surprising range of adaptations to life on sandy shores. Many of these adaptations are related to the instability of sand, and to the need for very rapid responses to changing conditions. Here we describe some of their adaptations, reserving until Chapter 4 the adaptations to more stable, finer-grained sediments.

Diatoms that exploit the surf zone

Most algae live either on the sediment surface (benthic) or float in the open water (planktonic). On relatively sheltered beaches there may be benthic coats of epipsammic diatoms or bubble-mats of cyanobacteria, like those described in Chapter 2. In the surf zone such a life would be impossible. Surprisingly, however, the surf zones of dissipative beaches in parts of Africa, Australia, and North and South America do have abundant diatom populations (Fig. 3.1). These diatoms move in a cycle between the sediment and the water surface, and their productivity can be very high—we examine their input to sandy-beach food webs later.

The diatom *Anaulus birostratus* provides a good example of how such a life is possible (Fig. 3.2). *Anaulus* accumulates at the water surface during the day, where it secretes mucus and may even produce a foam layer which helps floatation. Surface currents keep the foam near the shore. During the afternoon the diatoms drop out of the foam, and offshore rip currents take them to the outer edge of the surf zone where they enter the sand. Most cells remain in the substratum at night, where the mucus helps them to attach to sand grains. In the morning, they start to divide, lose the mucus coat, and float up to the surface where currents take them shorewards again. The cells thus undergo both a vertical migration and a horizontal onshore–offshore migration each day (Talbot *et al.*, 1990).

Invertebrate surfers

The surf zone is a dangerous place to be—but at the same time it is highly productive, and several quite different types of invertebrate are so well adapted to life there that they surf up and down the shore as the tides rise and fall (Brown *et al.*, 1989). The bivalve *Donax*, for example, surfs on incoming waves by extending the foot and the two siphons, which between them provide sufficient surface area to support the animal in the water. As the wave slows, *Donax* rapidly burrows and then suspension feeds on the prolific phytoplankton in the surf zone. In the Eastern Cape, South Africa, *Donax* makes up more than 90% of the benthic biomass.

Fig. 3.2 Schematic diagram showing how surf-zone diatoms cycle between floating in water and binding to sediment. Solid line shows floating, dashed line shows degree of mucus coating and dotted line shows number of dividing cells. (After Talbot *et al.*, 1990.)

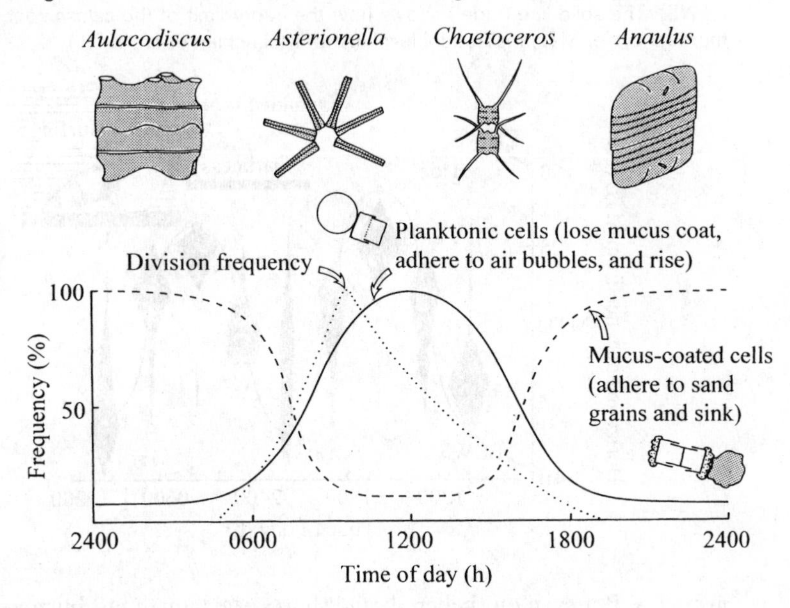

Another suspension feeder in the surf zone is the mole crab, *Emerita*. This genus has modified antennae that filter plankton but can then be rolled up when the animal migrates by surfing. The body is barrel-shaped and ideally adapted for rolling in the surf: as waves move up the shore, mole crabs emerge from the sand in large numbers and allow themselves to be rolled up the beach. They then dig rapidly and wait for the next wave. On the ebbing tide, they repeat the process in reverse, so that they follow the surf zone up and down.

Quite a different feeding mode in the surf zone is exploited by the snail *Bullia*, which makes a living by preying on animals stranded there, dead or alive. *Bullia digitalis*, in South Africa, lives mainly on coelenterates, but will also tackle *Donax*. *Bullia* surfs by expanding its foot and using it as an underwater sail. As the foot is slightly concave dorsally, it offers very great resistance to the water, and the snail surfs rapidly up and down—presumably a more economical way of traversing large distances than crawling. Populations can move up and down the shore as much as 50 m in one tidal cycle (Fig. 3.3).

Burrowing

Whether they can surf or not, all surf-zone dwellers need to be able to maintain their position on the beach. The only effective way for them to do this is to burrow, and burrowing is the most widespread adaptation found in the whole sandy beach macrofauna. Surf-zone dwellers have to be particularly fast—some *Donax* species can burrow completely in 5–6 s, while some *Emerita* species take

Fig. 3.3 Movement of the snail *Bullia rhodostoma* over an exposed sandy shore near Port Elizabeth, South Africa. The kites show percentages of snails found at positions between extreme high water mark of spring tides (EHWS) and extreme low water of spring tides (ELWS). The solid line ('tide') shows how the upper limit of the swash zone varies over the tidal cycle. MTL, mean tidal level. (After McLachlan *et al.*, 1979.)

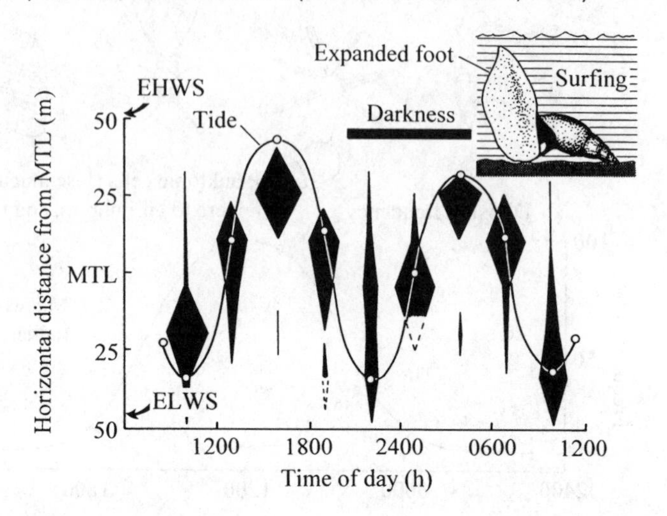

just 1.5 s. But even on sheltered sand shores, the fauna must burrow to prevent desiccation and to avoid contact with predators.

Soft-bodied animals inflate part of their body as an anchor, and then pull the rest downwards by contracting longitudinal muscles. Thus they usually can be seen to dig down in a series of discrete steps (Seymour, 1971). The lugworm *Arenicola* provides a good example (Fig. 3.4). Here burrowing starts when the proboscis probes gently into the sand while the anterior segments form flanges that act like ratchets, holding the body in place. This process of probing (stage I) continues until the head and two or three anterior segments are buried. The worm then forces body fluid powerfully into the proboscis to enlarge it so that the pharynx is everted. Proboscis and pharynx together form a 'dilation anchor'. The anterior segments form flanges again, take over as an anchor, and the proboscis is drawn back into the body. The relaxation of the proboscis draws water through the nearby sand and softens it, allowing elongation of the segments to propel the worm further downwards. This 'flange-anchorage' sequence is typical of stage II in the burrowing process. In stage III, the flanges are still formed, but instead of acting as an anchor, they help the proboscis as digging tools. Burrowing now therefore proceeds by a series of 'flange-proboscis' sequences in which the flanges dig while the proboscis forms an anchor. Using these mechanisms, the entire worm can disappear below the sand surface in less than 5 min.

Burrowing in bivalve molluscs, which do not have segments like polychaete worms, nevertheless shows the same principles. In *Donax*, for instance, the muscular foot probes into the sand, then swells to form an anchor. The shell

Fig. 3.4 Burrowing by the lugworm *Arenicola marina*. The solid line shows downward penetration into the sand in steps. The inset shows the worm performing one of these steps using the 'flange-proboscis sequence'. Arrows on the main diagram show when the burrowing sequences start—open arrows indicate flange-anchorage sequences, closed arrows indicate flange-proboscis sequences. (After Seymour, 1971.)

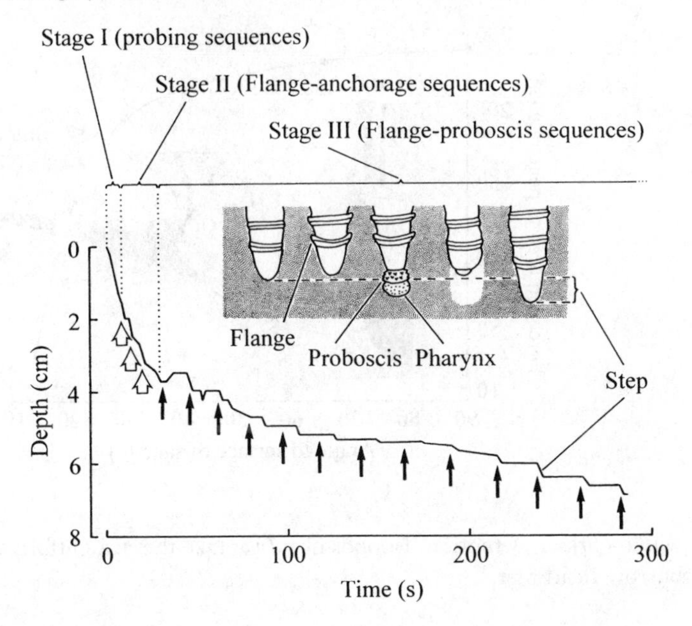

and body are then drawn down into the sand by retractor muscles in the foot. While the shell in turn forms a secondary anchor, the foot probes again and the sequence is rapidly repeated. In the water-saturated sand of the surf zone, gentle pressure exerted by the foot when probing liquefies the thixotropic sand and aids penetration, and jets of water from the mantle cavity aid this process. But because *Donax* burrows vertically, considerable penetration force is required (Fig. 3.5).

In contrast, the snail *Bullia* enters the sand at a very oblique angle, and a much lower force is needed (Fig. 3.5). Each step still consists of an alternate use of two anchors: after downward movement by the anterior part of the foot, this forms the first anchor, following which the body is pulled downwards. The posterior part of the foot then forms the second anchor while the anterior part moves downwards again.

Crustaceans have quite different mechanisms of burrowing. They have a hard exoskeleton and instead of changing the shape of the body they use their jointed appendages as digging tools. They show an enormous diversity of burrowing modes, and have the advantage over soft-bodied animals that they can easily burrow in dry sand. *Emerita* burrows backwards, using the tailpiece and the walking legs to scull into the sand. Most of the crabs, such as *Ocypode*, burrow sideways, while *Mictyris*, the soldier crab, appears to twist itself into the sand

Fig. 3.5 Resistance of the sand on a beach in South Africa to penetration, in relation to the angle at which the penetrating force is applied. Penetration was measured using a mechanical spring-loaded piston. The bivalve *Donax* usually burrows vertically and requires more energy to do so than the snail *Bullia* which burrows obliquely at an angle of 10–15°. (After Brown and Trueman, 1991.)

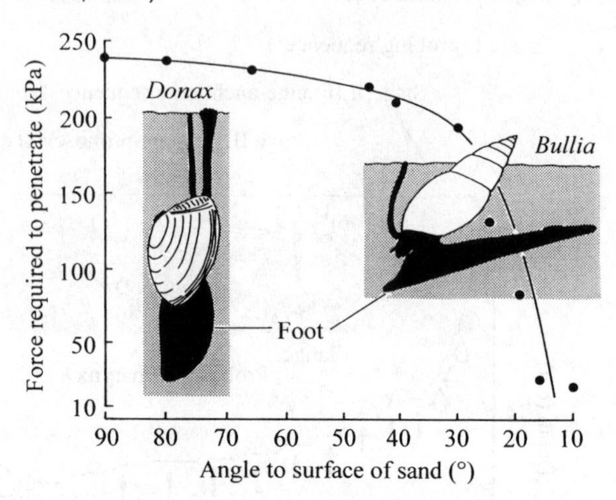

with a corkscrew motion. Isopods like *Tylos* take the straightforward option, and burrow head first.

Orienteering and rhythmic behaviour

Many burrowing species never move out of the sand unless disturbed by the waves. But some, particularly the smaller crustaceans, only use the sand as a refuge and feed on the sand surface or in the water column. This means that they need ways of orientating themselves, often in areas where the environment is relatively uniform so that there are few obvious cues to direction.

Eurydice pulchra is a predatory isopod that burrows in mid-shore sands at times of low tide, but swims in the water column searching for food when the tide rises. *Eurydice* has an endogenous (i.e. intrinsic) rhythm that triggers it to swim every 12.4 h. This period is approximately the same as that of the tides, so it is called 'circa-tidal'. It prepares the animals for swimming when the tide is rising, and ensures that they drop out of the water column and return to the sand *before* the tide falls, so they are not stranded above their normal levels on the beach. This mechanism sounds simple. But how does *Eurydice* cope with the more complex problems generated when the tidal amplitude changes from low, at neaps, to high at springs? Here another rhythm controls the degree to which the isopods emerge from the sand: at neaps, most stay buried, while on springs and especially on falling springs, the whole population emerges to swim. This ensures that the animals are not left behind high up on the beach where neap tides would not cover them. The rhythm peaks every 14 days, roughly half the time of a lunar cycle, so it is called a 'circa-semilunar' rhythm (Reid and Naylor, 1985).

Amphipods such as the European *Talitrus saltator* and the west-coast American species *Megalorchestia corniculata*, as well as the widespread isopod *Tylos*, have rather more complex orientational problems to face. They burrow near the high-water strand line in the day, emerging at night to feed on decaying seaweed. Some individuals may then travel long distances up into the sand dunes and down over the beach searching for food. On the often featureless sand flats, the problem of getting home to the right zone would challenge many human orienteers.

Talitrus and *Megalorchestia* appear to have two entirely separate mechanisms for finding their way about. Experiments in which the animals are placed in small circular chambers, so that their direction of orientation can be observed, have shown that they can navigate using the position of the sun as a 'sun-compass'. As the animals are primarily nocturnal, however, the actual use of this ability is somewhat obscure, although it is possible that oriented jumping may have evolved so that animals can rapidly escape from predators searching in the strand line (Ugolini, 1996). Similar experiments have shown an ability to navigate by the stars, so presumably some form of celestial navigation may occur in their normal lives. Other experimenters have chosen to observe animals released on the beach instead of in small chambers. These experiments clearly show that the amphipods have very good vision, and that they can orientate using local visual clues such as silhouettes to find their way up and down the beach (Hartwick, 1976). Which of these two mechanisms is used, or are they both employed?

Both mechanisms vary in their importance depending upon external conditions and upon the physiological states of the animals, and both are also under the control of endogenous rhythms. *Talitrus* and *Tylos*, for instance, have circadian rhythms (with a period of just over 24 h) which trigger activity during the night (Fig. 3.6). *Talitrus* also has a rhythm controlling the degree to which it will hop

Fig. 3.6 Activity of the amphipod *Talitrus saltator* in constant dim red light (equivalent to darkness because *Talitrus* cannot see red light). The activity is limited to the period during which it would be dark in the natural environment, showing that the rhythm is endogenous, with a period of about 24.5 h. (After Bregazzi and Naylor, 1972.)

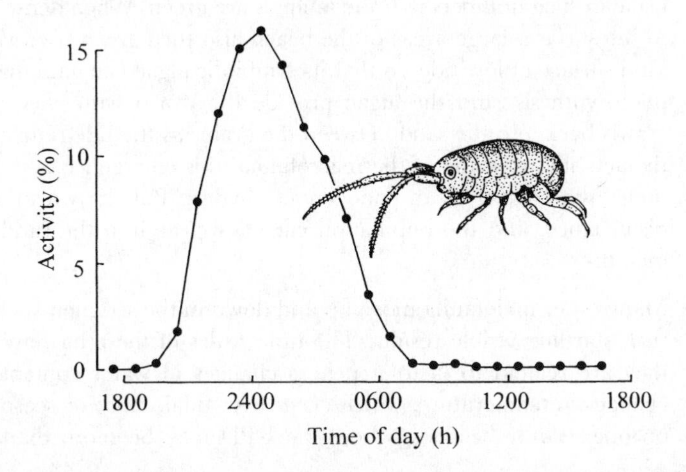

towards a light/dark boundary, and this rhythm switches on just before dawn. It is just at this time that the animals make for the home zone. The capabilities of the animals thus seem to be quite well understood. Because of the variability in the animals' responses to various cues, however, there is little agreement about the mechanisms they actually employ to get home in the real world.

How to be small: the ecology of meiofauna

The macrofauna—those animals that are retained on a sieve with a 0.5 or 1.0 mm mesh—can live within the sand only by making burrows, but for smaller animals a quite different lifestyle is possible. The meiofauna are metazoans that would nominally be retained on screens with a mesh size of 30–100 μm, but would pass through meshes of 0.5–1.0 mm. In practice, meiofauna are not usually sieved in this way because the sand grains (63–2000 μm in diameter) are larger than they are. Meiofauna are also very often long and thin, and can crawl or wriggle through mesh sizes that are very small in comparison to their length. Understandably, then, there can be no very clear distinction between 'small macrofauna' and 'large meiofauna'. Regardless of their shape, however, meiofauna on sand beaches are small enough to live between the sand grains, without forming specific burrows.

The meiofauna contain an extraordinary diversity of forms, from larval stages of larger species to taxa like the gastrotrichs and mystacocarids which are exclusively interstitial. The majority of meiofauna, however, belong to two taxa, the nematodes and the harpacticoid copepods. Along with oligochaetes, tardigrades, mites, ostracods, and others, they show remarkable adaptations for adhering to sand grains and wriggling between them, but are seldom seen by the average shore biologist. Instead, we begin by turning to an animal that can make a visible impression on the beach—a small turbellarian worm, *Convoluta*. *Convoluta roscoffensis* is abundant on the Channel coast of France, has been recorded in the Severn estuary in England, and has relatives in South Africa. Most interstitial turbellarians are predators, but *Convoluta* contains symbiotic algae in such numbers that the animals are green. When dense populations rise to the surface, large areas of the beach also turn green. *Convoluta* comes to the sand surface at low tide, so that its symbiotic algae can gain maximum light for photosynthesis, and the algae provide the worm with energy. *Convoluta* then crawls back into the sand between the grains as the tide returns (Fig. 3.7). Like the activities of some of the macrofauna, this emergence and disappearance is under the control of an endogenous rhythm. But the worm also responds to disturbance, and the population can disappear into the sand as a protective measure.

Many other meiofauna move up and down in the sediments, although not with such startling visible results. The time scales of these movements suggest that they are related to factors such as changes in water content of the sand or changes in temperature, which occur over tidal, daily, or seasonal cycles. Thus on open sandy beaches, where the RPD may be more than 1 m below the

Fig. 3.7 Vertical movement of the small flatworm *Convoluta roscoffensis* in relation to tidal cycles. The thick black layer shows position of the worms. The thin line shows tidal rise and fall. In the upper diagram, low tides occur during the day and at these times the worms move to the sand surface. In the lower diagram, one low tide occurs at night and the worms then stay buried in the sand. HW shows high water, LW shows low water. (After Gamble and Keeble, 1903.)

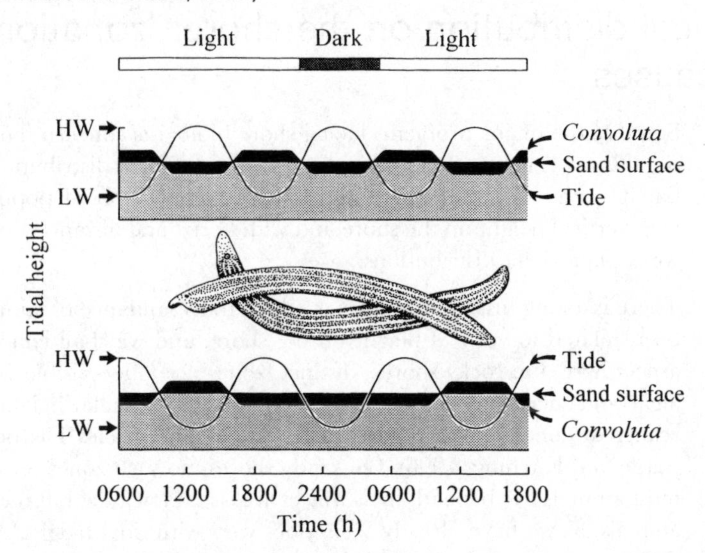

surface, nematodes and harpacticoid copepods move upward in the sand as the tide rises, and then may descend 10 cm or more when the sand dries out. On sheltered beaches, where the RPD is nearer the surface, the distribution of many meiofauna appears to be related to burrowing macrofauna, as we noted in Chapter 2. This is at least partly because the macrofauna may increase oxygen levels and provide nutrients, but may also be related to changes in sediment structure.

One of the major questions in meiofaunal ecology concerns the importance of other interactions between meiofauna and macrofauna. Do macrofauna eat meiofauna? And do meiofauna therefore provide an energy input into the major food webs on sandy beaches? As meiofauna may have population densities of the order of 10 million/m^2, and meiofaunal biomass can sometimes exceed that of the macrofauna, this is evidently an important question in terms of trophic dynamics.

On exposed shores, the general consensus is that meiofauna form an essentially 'closed' community which has its own producers, herbivores, detritivores, and carnivores, but does not interact to any degree with the macrofauna. But in more sheltered areas, deposit feeders such as the lugworm *Arenicola* may actually 'garden' meiofauna and consume them. Furthermore, experiments on the tidal sand flats of the Wadden Sea, north Germany, have shown that meiofauna are attacked by macrofaunal predators at high tide: crabs like *Carcinus* and shrimps like *Crangon* sift through the surface 10 mm of sediment, consuming harpacticoid copepods and nematodes, while gobies (*Pomatoschistus*) feed on the surface. When these predators are excluded, meiofaunal numbers increase dramatically. There is, of course, also the possibility that meiofauna may be involved in food chains indirectly, and we shall consider this aspect on p. 54.

Vertical distribution on the shore: 'zonation' and its causes

Some of the major problems facing shore biologists concern the distribution of organisms. Plants and animals never show a uniform distribution, but occur in patches, and the abundance and species composition of populations change with vertical height on the shore and with horizontal distance along it. How can we explain these distributions?

There is a long history of studies in which the abundance of fauna and flora has been related to vertical height on the shore, and we shall concentrate on this aspect here. On rocky shores, distinct 'zones' are often visible where species of algae and encrusting animals come to dominate particular heights, although the reality of zones for more than a few dominant species has been questioned (Little and Kitching, 1996). On sandy shores, no such zones are visible because most animals live beneath the surface, at least at low tide; but the distribution of animals, as we have already seen, does vary with tidal height. Amphipods like *Talitrus* occur very high on the beach while surf-zone suspension feeders like *Donax* and *Emerita* are found lower down.

Early studies concentrated upon the influence of physical factors on such patterns. For example, where particle size changes grossly with vertical height, as it may do on estuarine beaches, the fauna changes abruptly too. Thus in the Severn estuary, England, amphipods such as *Bathyporeia* and *Corophium arenarium* occur at the top of the shore in clean sand, but suddenly disappear and are replaced by polychaetes such as *Nereis* and *Nephtys*, and bivalves like *Macoma*, where the substratum changes to muddy sand (Boyden and Little, 1973). More subtle effects related to changes in particle size are therefore to be expected on clean sand beaches, although as discussed in Chapter 2, the effects may be very indirect.

In particular, particle size will affect how much the substratum dries out at low tide (see p. 15), and as few invertebrates can resist desiccation, drying may be an important controlling factor. Not surprisingly, then, some authors who have attempted to define 'zones' on sandy shores have based these on degrees to which the sand retains moisture—upper zones will have dry or damp sand, while lower ones will remain saturated. Other authors have chosen to use species as indicators, instead of physical marks: there is often an upper zone of ocypodid crabs and talitrid amphipods, a midshore zone with isopods and a low-shore zone characterized by mole crabs, amphipods, and bivalves. Although based on very different premises, these schemes define very similar zones, and a recent scheme has incorporated features of both (McLachlan and Jaramillo, 1995).

The characteristics of zones on sandy shores change with gradations between flat dissipative beaches and steeper reflective ones (Fig. 3.8). On both, there is a 'supralittoral zone' in which the sand is dry and where ocypodid crabs dominate in the tropics and talitrid amphipods are found in temperate zones. The isopod *Tylos* may be found there, and the zone grades into the more terrestrial areas of the sand dunes. Below is a region of damp sand, the 'littoral zone', in which amphipods like *Haustorius*, carnivorous isopods like *Eurydice* and spionid polychaetes are common. This zone is predominant on reflective shores, but may be very narrow in the flatter dissipative regions. Lower down again is the region where the water table comes to the surface, so that water may actually flow out of the sand when the tide is low. The 'effluent line' has been called a 'zone of resurgence' by some authors, and marks the top of the region where the sand remains saturated. This zone, the 'sublittoral', is characterized on dissipative shores by predatory polychaetes such as *Nephtys* and *Glycera*, and suspension feeders like the bivalve *Donax* and the mole crab *Emerita*, but may have very little fauna on steep reflective beaches.

While such a classificatory scheme is useful, it is not always easy to define exact limits to the proposed zones. Indeed, zones on sandy shores are in general much less sharply defined than those on rock. Why is this? One reason is that many animals living in sediments are not fixed in position, and tend to move up and down in cycles that follow tides or seasons rather than staying in one place. Patterns of zonation observed on one occasion only may not therefore give a really representative picture. Another reason relates to the three-dimensional nature of sedimentary environments: there is room for several species to live at any one level on the beach without interfering with each other, something that would be much more difficult on rocky shores. Distributions of species on sandy shores therefore often overlap widely.

Fig. 3.8 Scheme of zonation on sandy shores, showing changes from dissipative to reflective beaches. (After McLachlan and Jaramillo, 1995.)

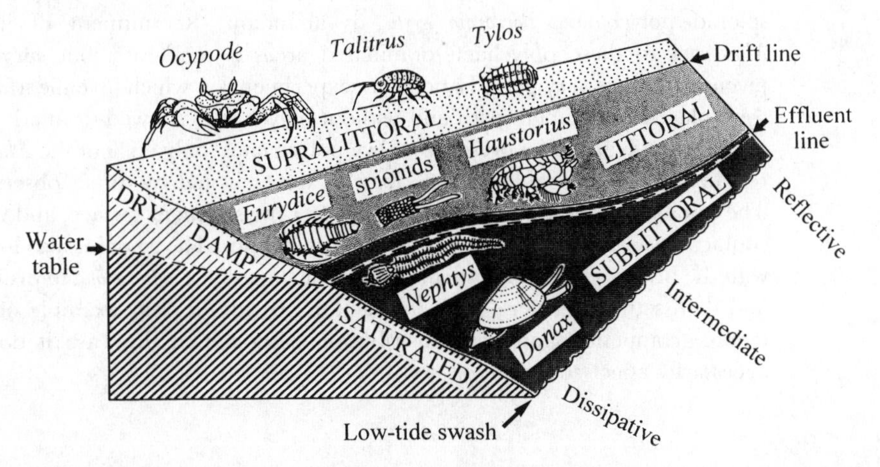

As the zonation scheme just described is heavily weighted in terms of one physical factor, water content of the sand, it is important to ask whether biological interactions are also important in creating and maintaining zones. On rocky shores, factors such as competition, predation, and larval settlement are known to be particularly relevant here (Little and Kitching, 1996). On sedimentary shores, we have already noted the interactions between macrofauna and meiofauna, and between deposit feeders and suspension feeders, many of which act via bioturbation and disturbance (Chapter 2). But do these effects merely cause patchiness, or do they influence vertical zonation? And while they are well known on sheltered shores, do they also occur on exposed sand?

Observations of isopods on exposed sandy shores suggest that physical factors are not the only forces structuring zonation, and that competition is one 'biological' factor that may be a potent force determining zonation patterns (Defeo *et al.*, 1997). In Uruguay, two species of the isopod *Excirolana* are common. When both species occur on the same beach, *E. braziliensis* is usually restricted to high levels while *E. armata* is found lower down. But when *E. brasiliensis* is the only species on the beach, it spreads down to the swash line and thus occupies the zone normally taken by *E. armata.* These distributions suggest that *E. armata* normally outcompetes *E. brasiliensis* low on the shore, and excludes it. The fact that individuals of both species, collected in the field, are smaller where the two species occur together, supports the idea of interspecific competition. Other explanations based on preference for particular sizes are possible, but are not supported by laboratory experiments. On the other hand, laboratory experiments do not show a significant effect of *E. armata* on the survival of *E. brasiliensis* over periods of 4 weeks, so effects of competition must be quite gradual.

Some experiments on a sheltered sandy shore give more conclusive evidence that one species of infauna can effectively oust another (Cummings *et al.*, 1996). In Manukau Harbour, New Zealand, the dominant bivalve is *Macomona liliana*, but in some areas this occurs at low densities. In these areas, a tube-building spionid polychaete, *Boccardia syrtis*, is abundant. Recruitment of juvenile *Macomona* to these polychaete-dominated areas is moderate, but survival of juveniles there is very poor. Laboratory experiments in which juvenile *Macomona* were placed in sediment with and without polychaetes showed dramatic differences in survival: in the presence of live polychaetes, only 25% of the *Macomona* remained. The probable reasons for this were elicited by direct observation. The spionid *Boccardia* waves its two palps over the sediment surface, and when it contacts *Macomona*, the bivalve withdraws its siphons. The spionids also interfere with the bivalves by picking them up. Both actions cause *Macomona* to drift away, and then settle elsewhere. This interaction provides a clear example of interference competition affecting distribution, although in this case it does not necessarily affect zonation.

Biological interactions

The discussion of zonation on the shore has emphasized that organisms do not live in isolation. Herbivores need to find a source of plants, carnivores need prey, and even detritivores need the remains of other organisms for food. But there are many other interactions that cannot be so neatly pigeon-holed. For example, in Chapter 1 the importance of many of the burrowing macrofauna for meiofauna was discussed, as they influence conditions in the sediments. In the past, meiofaunal food webs have been thought of as self-contained, not having interactions with the macrofauna. But is this really true? Until recently, too, sandy shores have been considered prime examples of physical control by a harsh environment, where processes such as competition have little impact. As we shall see later, such views are changing.

Commensals and other associations

The simplest way to become aware of interactions between species on sandy shores is to investigate associations in the field. Many macrofauna in sand are accompanied by other species, either attached to them, or living in the same burrow, or just living nearby. These relationships may be beneficial to both partners (mutualism), to one only but with no harm done to the other (commensalism), or beneficial to one but detrimental to the other (parasitism). As yet the degree of benefit to each partner has seldom been defined clearly on sandy shores, but we will use the term 'commensal' for most interactions.

In the surf zone, commensals would hardly be predicted because animals seldom have fixed burrows. Nevertheless, the bivalve *Donax* and the predatory snail *Olivella* in southern California often have colonies of a hydroid, *Clytia*, attached to their shells. When the hosts are buried in sand, the polyps of the hydroid project into the water for feeding, and have the advantage of being attached to the only solid surface in the region.

When conditions are less fierce, and the sand has some finer components, sipunculid worms such as *Golfingia* burrow and feed. Very often these worms have commensals attached to their posterior end, taking advantage of the respiratory currents produced by the worm. These commensals are entoprocts, a phylum of very small suspension feeders. Similarly, some of the burrowing sea-cucumbers have commensal bivalves attached.

The majority of commensals live in the host's burrow rather than being directly attached. Classic examples are found in the burrow of the shrimp, *Callianassa* (Fig. 3.9). Some members of this genus are found in clean sand, but most occur where mud is mixed in, and there they produce burrows which are roughly U-shaped, but may have several openings and blind tunnels. Within the burrow there may be up to five different species of commensal at any one time, and *Callianassa californiensis* plays host to at least 10 species overall. Most commonly, it shares the burrow with a goby, *Clevelandia*; a polychaete scale worm, *Hesperonoë*;

Fig. 3.9 The burrow of the ghost shrimp *Callianassa californiensis* showing some of its commensals: the scaleworm polychaete *Hesperonoë adventor*; the pea crab *Scleroplax granulata*; the goby *Clevelandia ios*; and the bivalve *Cryptomya californica*. (After MacGinitie and MacGinitie, 1949.)

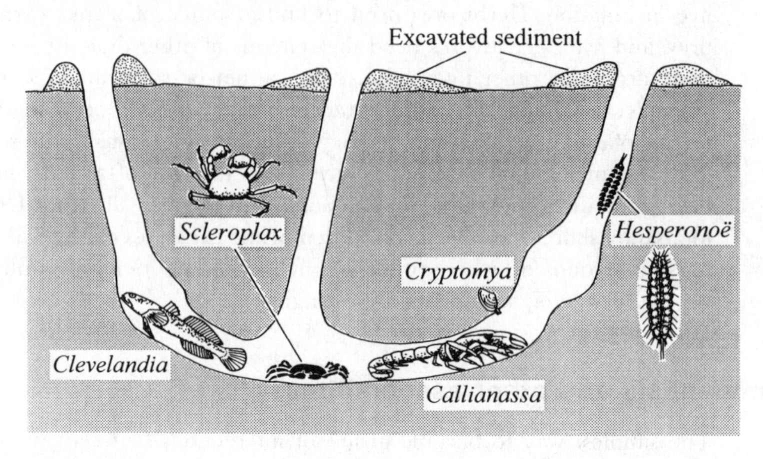

and pea crabs, *Scleroplax*. In addition, a small clam, *Cryptomya*, often burrows into the sand nearby and its siphons project into the burrow of the shrimp. The commensals presumably derive shelter from the burrow, and some may steal food from the host.

These and other surprising associations (MacGinitie and MacGinitie, 1949) make it quite clear that beneath the sand surface, a whole range of interactions occurs, varying from some degree of collaboration to competition for resources, predation, and parasitism.

Competition

Despite the difficulties of working in an environment where most organisms are out of sight for most of the time, many observers have shown that competitive interactions in sand are widespread (Wilson, 1991). On exposed shores, competitive effects may be slight, but in shelter species may interact either directly or more often indirectly, for instance by building tubes or by disturbing the sediment.

On sandflats in southern California, one of the common bivalves is *Sanguinolaria nuttallii*. When this species is experimentally confined with two other deep-burrowing species, its growth rate declines by 80%. This effect might be due to competition for food or to competition for space—factors that are very hard to differentiate in an environment where food arrives in suspension or is scattered between the sediment particles. But in this case, when the species in competition with *Sanguinolaria* are replaced by the valves of dead specimens, the growth of *Sanguinolaria* still declines, showing that the species are competing not for food but for some aspect of available space. *Sanguinolaria* also competes with other species such as the burrowing shrimp, *Callianassa*; when the shrimps are removed, numbers of *Sanguinolaria* increase. In fact *Callianassa* has a major influence on sand structure because of its burrowing activities, and when it is present many other species are reduced in numbers.

While interactions between adults are thus common, interactions between adults and larvae may be of even greater significance. Various species of amphipods, polychaetes, and gastropods have been shown to affect the settlement and survival of other species. The suppression of larval settlement, or reduction of post-larval survival, may have a powerful influence on community structure, and we discuss this phenomenon in more detail in Chapter 6.

Competitive effects can only be convincingly demonstrated by experiment, and on exposed sandy shores these are virtually impossible. But one of the important points to note here is that whether the experiments are attempted in the field or in the laboratory, they are inevitably on a very small scale relative to the area over which it is postulated that competition occurs. Does this mean that competition occurs in the real world, or could some species avoid competition by moving away, or by finding temporal refuges, or by some other process? The whole problem of how biological processes that operate over wide areas can be assessed and interpreted from small-scale experiments is a current focus of interest of benthic marine ecologists (Thrush *et al.*, 1997).

Food capture: the suspension feeders

The abundance of surf-zone diatoms on some beaches ensures a supply of food for those specialist suspension feeders that can tolerate surf conditions. These include many mobile crustaceans, particularly prawns and mysids, and sometimes amphipods. In many cases swarms of these animals migrate into the surf zone to feed and then move offshore, sometimes resting on the bottom. They themselves are subject to predation by fish that move into the surf zone for short periods. Juveniles of such genera as mullet, *Liza*, can vary their diet from the surf-zone diatoms themselves to the zooplankton that feed on them, so it is hard to classify them in any one trophic level.

The supply of suspended food is in most areas extremely variable (Okamura, 1990), so most suspension feeders show great behavioural plasticity. Indeed, the major characteristics of suspension feeders on exposed sandy shores are adaptability and mobility. The surf-clam *Donax*, for example (Fig. 3.10), can filter a wide range of particle sizes, can utilize both phytoplanktonic algae and plant detritus, and can even assimilate the mucus 'foam' produced by the surf-zone diatoms with an efficiency of 50%. The mole crab *Emerita* feeds just as the waves wash down the beach by digging rapidly into the sand and then spreading its feathery second antennae just above the sand surface to strain out food particles. Meanwhile its first antennae form a sort of vertical tube acting as a respiratory siphon (Fig. 3.10).

Above the surf zone, many species use suspension feeding to trap fine particles, but also pick up coarse particles from the substratum—a form of deposit

Fig. 3.10 Three different mechanisms of suspension feeding. The mole crab *Emerita* strains water through a setal array on the second antennae. The bivalve *Donax* uses gills and palps. The spionid polychaete *Paraprionospio* uses two palps with ciliated grooves (only one shown here) to take sediment to its everted pharynx. (After Ruppert and Barnes, 1994, Ansell, 1981, and Dauer, 1985.)

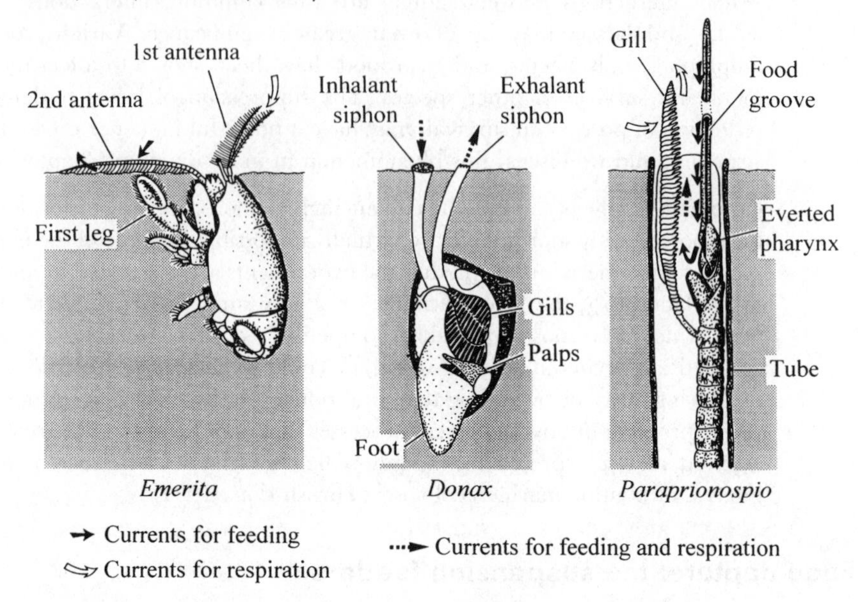

feeding. Burrowing amphipods, such as *Haustorius*, alternate between these two modes depending upon supply: large masses are picked up directly by the mouthparts while finer particles are taken using complicated arrays of setae on the mouthparts as filters. A similar diversity of feeding can be found in polychaetes like *Spio*. Spionids make sandy tubes, and from these project two long ciliated palps which have ciliated grooves running along them. The palps can be used to pick up detrital food from the sediment surface, but can also trap suspended food (Fig. 3.10).

As exposed sandy shores grade into more sheltered areas with a higher mud content, and greater stability, more benthic algal food in the form of epipsammic diatoms is available, and there is more detritus. There are many types of suspension feeders in these habitats: polychaetes such as *Sabella*, with a complex crown of filtering tentacles; bivalves, such as *Mercenaria*, and the cockles *Cardium* and *Cerastoderma*, with short siphons and capable of producing strong filtering currents; burrowing anemones and a variety of burrowing snails. But the most noticeable change is towards an abundance of deposit feeders: annelids, such as *Clymenella* and *Arenicola*, the shrimp *Callianassa*, and a horde of long-siphoned bivalves. Discussion of deposit feeding is postponed to Chapter 4, as it reaches its peak in mudflats rather than in sand.

Food capture: the predators and scavengers

With a high diversity of suspension feeders making a living out of phytoplankton, and with the possibility of oceanic animals being stranded, sandy beaches offer a wealth of opportunity for predators and scavengers. But the carnivorous life also poses severe problems: stranded food is probably even more patchy in time and space, and therefore more unpredictable, than suspended food; while most suspension feeding prey is buried in the sand and not easy to detect. In response, carnivores on sandy beaches can be seen to be opportunists, to have developed more than one mode of feeding, to be able to ingest huge meals at one go, and then to be able to fast for long periods.

Perhaps the most striking example of these approaches is given by the South African snail *Bullia*. Its main diet consists of coelenterates—jellyfish and siphonophores washed up on the shore. It detects these with its very sensitive chemoreceptor, then crawls towards the prey in the swash zone. Here the food may be rolled about by the waves, so the snail anchors itself to the coelenterate by thrusting its proboscis deeply into the prey's flesh. *Bullia* can then consume an amazing volume of food in one sitting: up to one-third of its own tissue weight in 10 min! It can then forego feeding for more than 10 days. If carrion is not available, *Bullia* will attack living prey like the surf-clam, *Donax*. In addition, some *Bullia* species have algae that grow on the apex of the shell, and with its long proboscis the snail can crop the alga periodically—an instance of algal gardening very like that seen in rocky-shore snails. *Bullia* is thus well adapted to the unpredictable and variable nature of food in the surf zone.

Other snails such as the moon snails, *Polinices* (formerly called *Natica* in Europe) are found worldwide on sandy beaches and tend to be more consistently predators. *Polinices* has a globular-shaped shell and a wedge-shaped foot that can be vastly expanded by taking in sea water to produce both a ploughing organ and a streamlined cover for the shell. It ploughs along just below the sand surface, and when it makes contact with bivalve prey it clasps the prey's shell firmly with the foot, drills a neat hole, and inserts its proboscis. The flesh can then be rasped away and sucked out. *Polinices* is often accompanied by carnivorous whelks like *Oliva* and *Olivella* which are streamlined for burrowing in the sand.

Predators that are in the main confined to the surface of the sand either wait for prey to pass by, or have to dig or probe for food. There is a great variety of these surface predators, from birds and fish to crabs and, in some areas, insects. Insects are often ignored by marine biologists, but in the Chesapeake Bay the beach tiger beetle, *Cicindela dorsalis*, attacks amphipods. Adults hunt directly, but larvae adopt a sit-and-wait policy. They live in vertical burrows, using their head to form the 'roof', and ready to seize passing insects or amphipods. In the tropics and subtropics, ghost crabs, *Ocypode*, are opportunist feeders: they search actively for prey and often dig into the sand to reach it; but they also scavenge, and they can use their mouthparts to sift though sand, separating out meiofauna.

Vertebrate predators share out the beach over the tidal cycle. At low tide, gulls search for food left stranded, while waders probe into the sand. The sanderling, *Calidris alba*, is a good example of a wader that feeds in the surf zone: it dashes down the beach as the waves run out, probes in the sand for amphipods, worms, or molluscs, then runs back up the beach as the next wave rolls in. Other birds, such as oystercatchers, *Haematopus*, also probe into the sand, and in South Africa *Haematopus* actually detaches and eats the siphons of the surf-clam *Donax*. When the tide rises, many fish move upshore, feeding as they go. On sheltered shores, juveniles are often particularly common. They use the shore as a nursery area, as there is abundant food. Flatfish, for example, mature on the sheltered beaches of Oregon, USA, and those of northwest Europe. Plaice, *Pleuronectes platessa*, and dab, *Limandia limandia*, move into the shallows in summer, when only 2 cm long. They feed both on the demersal plankton (plankton near the bottom) and on the benthos. Benthic prey may include small macrofauna and meiofauna, and both flatfish may also feed on larger bivalves by nipping off the ends of their siphons.

Food webs and energy flow

By examining the feeding habits of individual species, or of guilds—that is to say, groups of species that feed in the same way—it is possible to build up a picture of how food webs are organized on sandy shores. On exposed, dissipative beaches, food webs begin with primary production by the surf-zone diatoms. Major herbivores are bivalves like *Donax* and the dense populations of mysids that shoal in surf-zone waters. Scavenging snails, such as *Bullia*, predatory fish, and birds form the next trophic level. This qualitative shore-bound approach has some disadvantages, however, because it leaves unanswered several questions. How is the sandy beach ecosystem connected with offshore, oceanic systems? If the interstitial fauna really forms its own circumscribed ecosystem, as suggested above, how does it obtain its energy source? How important are the various species in the ecosystem, and how does this importance vary from beach to beach?

One way to answer some of these questions is to construct a model of energy flow through the system. This is immensely difficult because it involves integrating enormous numbers of biological interactions over an area of beach and over not just tidal periods but over seasons. Nevertheless, it has been done, at varying levels of detail, for a number of sandy beaches.

Figure 3.11 shows a summary of energy flow on an exposed intermediate/dissipative beach on the Eastern Cape, South Africa. Here the only significant primary production is that of the surf-zone diatoms, but surprisingly only a small fraction of this production is consumed directly: zooplankton such as mysids, and benthic suspension feeders such as *Donax* consume approximately

Fig. 3.11 Energy flow on Sundays Beach, an exposed sandy beach on the south-east coast of South Africa. Units are given in mg C/m^2 per day. To simplify the diagram, respiratory losses are not shown, and figures are rounded. POC, particulate organic carbon; DOC, dissolved organic carbon. (After Brown and McLachlan, 1990, and Heymans and McLachlan, 1996.)

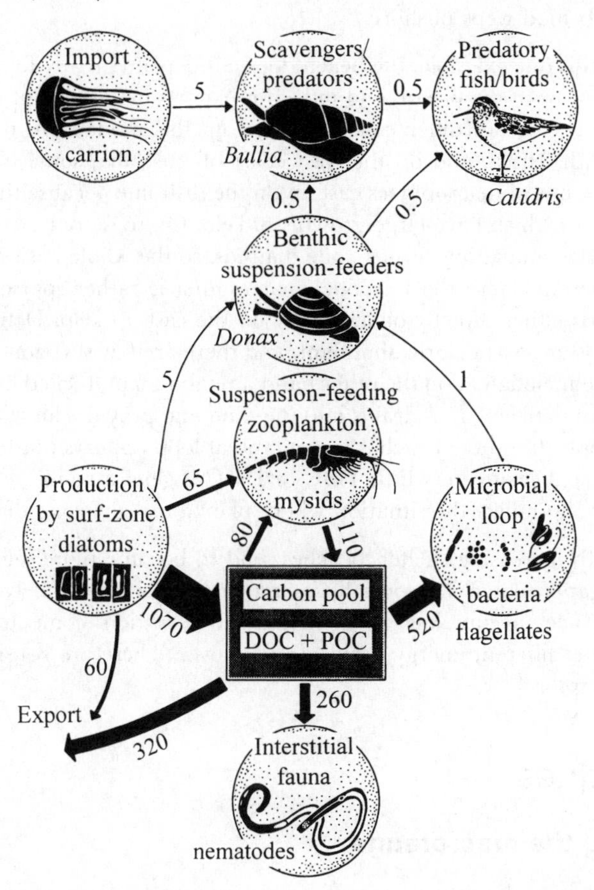

7% (Heymans and McLachlan, 1996). The remainder forms a pool of organic carbon, both particulate (POC) and dissolved (DOC), and this pool fuels two separate parts of the shore ecosystem. First, it fuels the interstitial fauna because water flushes through the beach sand as tides rise and fall, and the interstitial bacteria can utilize both the POC and the DOC that is carried with it. The protozoans and the meiofauna can then feed on the bacteria as well as utilizing some of the POC directly. Secondly, the carbon pool fuels the 'microbial loop' described in Chapter 2: bacteria on suspended particles utilize both POC and DOC, and these provide a food source for heterotrophic flagellates. The importance of these flagellates and of ciliates that feed on them in fuelling suspension feeders is not certain, but suspension feeders, such as *Donax*, can certainly filter out some flagellate-sized particles.

Transfer of carbon to higher trophic levels—predatory fish and birds—is very small, but there is also some input to carnivores from stranded carrion. In contrast to this, a sizeable fraction of the carbon pool is exported from the system, presumably as detritus that either ends up on more sheltered shores, or fuels food webs offshore.

In this one example, the beach does not depend greatly for its energy supply on input from the ocean, and indeed it exports material. But many other exposed sandy beaches differ considerably from this model. On the Western Cape in South Africa, and on the west coast of Australia, some beaches receive large amounts of macrophytes cast up on the drift line because they are near to rocky shores which have large growths of kelp. On these reflective beaches, there are no accumulations of surf-zone diatoms, so the whole basis of the energy flow is different. Here the intertidal macrofauna is rather sparse, and what there is feeds either directly or indirectly on the cast-up kelp. Detritus-feeding amphipods are particularly abundant, and there are few surf-zone suspension feeders. The meiofauna, on the other hand, are abundant, fuelled by the organic import from the kelp. This drains into the sand and provides an instant interstitial food supply. On these beaches the microbial loop appears not to be of great importance, because very little POC or DOC is generated in the water column. This is quite unlike the situation where planktonic surf-zone diatoms predominate.

With increasing shelter, beaches tend to become more and more importers of organic material, especially as much POC is very finely divided, and small particles accumulate in shelter. Deposit feeders come to predominate, and quite different energy pathways take over. These are considered at the end of Chapter 4.

Techniques

Sampling the macrofauna

Assessing the distribution of animals on sandy shores is far more difficult than on rocky shores, where the organisms are at least visible to the observer. Basic mechanisms for collecting the infauna are described by Holme and McIntyre (1984). Most sampling in intertidal sand is carried out using corers, which are pressed into the sand, closed with a plate at the bottom, and then lifted out. The sediment is then sieved to extract the macrofauna. Although this procedure sounds simple, it poses a number of problems. How large and how deep should the cores be? What mesh size will retain most animals while letting the sand particles through? How can one ensure that animals retained in the sieve are not missed when sorting? How should the cores be positioned on the beach to give an accurate impression of the distribution of the fauna?

Some of these points are best investigated in a pilot study before any large-scale work is undertaken, as there is no hard-and-fast rule for all beaches (James and Fairweather, 1996). For example, tests should be made with a range of corer sizes, say 50–200 mm or larger, to see where the balance lies between consistency and ease of use. Small cores will give more variable densities, but if small-scale variation is to be investigated, they may be appropriate. Large cores are clumsy and time-consuming to use, and they obscure small-scale variation, but may be necessary if some of the individuals to be sampled are themselves large. For larger, rarer animals, it may be necessary to use quadrats of, say 25 × 25 cm, instead of cores. The appropriate depth of cores, usually 10 or 20 cm, can be investigated in the same pilot study, as can the effects of mesh size. Although 0.5 mm is the standard size to separate macrofauna from meiofauna, it may be necessary to use 1 mm or even 2 mm mesh if the sand is coarse, because the particles block the sieve.

Animals retained by the sieve are usually easily visible, but if they are particularly transparent species, staining with Rose Bengal may help. Some workers preserve the fauna with formalin before sorting, and this has the benefit of preventing smaller forms from crawling through the mesh—on the other hand, many soft-bodied species become unrecognizable when pickled!

The decision about how to position the corers in a logical sampling array is perhaps most difficult, but most crucial, of all. In order to be meaningful, samples must be replicated, and because of the patchiness of most species, this replication must be considered on several scales. For example, a single transect down the beach, with single cores at regular intervals, will ignore the fact that at each height, community structure will be patchy, and also that the community may be different some distance along the beach. The best compromise is to have what is called a 'stratified' design to a sampling programme, in which groups of samples are taken at each point on a transect, while the transect itself is repeated some distance away. An excellent discussion of possible approaches is given by James and Fairweather (1996).

Sampling the meiofauna

The corers most frequently used to sample meiofauna are perspex tubes 30 cm long and 3.6 cm in diameter—chosen to give a cross-sectional area of 10 cm^2 (Giere, 1993). These can be pressed into the sand, and the contents can be extruded with a piston. The sediment can thus be sampled at intervals so that the fauna can be closely related to depth.

Extracting meiofauna, by definition less than 0.5 mm in size, from sands which may consist of particles greater than 1 mm in size, cannot be done by sieving. Instead, most techniques use differences in the specific gravity of organisms and mineral particles to separate them. Animals can often be decanted from the surface of the water after stirring, because they stay in suspension longer than mineral particles, but this extraction is seldom complete. Alternatively, a jet of water can be used to bring animals to the surface so that they can be creamed off—a process known as elutriation. This technique *can* be made quantitative.

4 A fine option: life on mudflats and in seagrass beds

There is no strict demarcation in particle size between sand beaches and mudflats. Despite the powerful sorting capacity of currents derived from waves and tides, both invariably have a wide range of particle sizes. In estuaries especially, sediment may even have alternating layers of mud and sand laid down by successive current regimes. Nevertheless, the differences between a clean sandy beach facing the open ocean (where particles larger than 63 μm predominate) and a sheltered mudflat (where many particles are less than 63 μm) are enormous. Unless artificially truncated by a sea wall, mudflats are usually fringed at the landward side by wide reaches of salt-tolerant vegetation growing below the high-water mark: salt marshes in the temperate zone, mangrove swamps in the tropics. They often have a subtidal fringe of vegetation as well, where the seagrasses like *Zostera* grow. Between these two vegetated zones, the muddy substratum has a high water content and a low permeability, and very often contains much organic material so that it is anoxic just below the surface. Mud particles are small in comparison with sand, but as we shall see they are less mobile, and so are more likely to remain in one place than sand. Conditions for life in mud are therefore very different from those in clean sand.

The majority of mudflats are found where the coast is sheltered from wave action, either because of offshore bars, or because an estuary or bay forms an indentation in the coastline. Most sheltered areas have some input of fresh water, and mudflats are classically associated with low salinity—the majority occur in estuaries. As invertebrate diversity appears to decline with decreasing salinity, one of the major problems for anyone running a marine fieldcourse is to find a high-salinity mudflat. Although at this stage it is important to note that there is much discussion over whether low salinity is the prime cause of low diversity, consideration of this topic is deferred to Chapter 8.

Before considering the biology of mudflats, here is a brief introduction to some of their physical characteristics.

Some physical features of mudflats

Mudflat morphology

Mudflats are not necessarily flatter than sandy beaches. True, they can have slopes as gentle as 1 : 1000, but mud can also form steep banks, especially where creeks cut into the flat, and steep erosion cliffs can form under unusual conditions of wave action or undercutting by tidal currents. Nor do mudflats consist entirely of mud—there is usually a substantial sand component. Mudflats usually occur where there is shelter from major attack by waves, but often where tidal currents are high. Sandflats, in contrast, are more characteristic of wave-dominated environments. The contrast is nicely seen when comparing the distribution of sediments on sandy beaches and mudflats. On sand beaches, most sediment is moved by waves or by the currents associated with them: heavier particles are moved upshore by strong incoming waves and the weaker backwash fails to take them downshore again, so coarser sediments usually accumulate high up on the beach and finer ones lower down (see p. 7). On mudflats, however, tidal currents are responsible for most movement of sediment. Tidal velocities vary over the tidal cycle and are highest at mid tide, falling to zero when the tide is high. At mean tidal levels (MTL), therefore, sediments are relatively coarse, while the finest silts tend to settle out high on the shore—quite the reverse of the situation where wave action dominates.

Mudflats also show a change in profile near MTL. Above this, where fine silts predominate, they are often fairly flat. Below MTL the sandier mud may be much steeper. But the most prominent feature of mudflats is the branching creek system which serves both to drain water off the flat as the tide falls, and to funnel the rising tide until it spills out over the creek banks. Water velocities within the creeks rise much higher than those on the flat itself, so creeks are often lined with coarse sediments while very fine silts are deposited near to them when the water flows on to the flats.

Fine sediments and the phenomenon of cohesiveness

The presence of fine particles, particularly those less than about 2 μm in diameter, has a profound effect upon the sediment as a habitat, because such particles are not simply held down by gravitational forces, but are, as discussed in Chapter 2, cohesive. As little as 5% of clay particles (by weight) produces some cohesion, so the small particles have an effect out of all proportion to their abundance, binding together the sediment.

Why are small particles cohesive? Particles less than about 63 μm in diameter are not generally solid individual grains like the quartz grains that make up the majority of sandy beaches. They are 'secondary' particles, made up of very small components held together by their own electrostatic forces. In fresh water, the surface charges on small particles cause mutual repulsion, but in sea water these charges are reduced, and the particles stick together to form 'flocs'. The importance of flocs in the formation of mudflats can easily be seen by comparing settling rates of different-sized particles in water: a 5 μm particle would fall approximately 7 cm in 1 h, while a 500 μm particle would fall 1800 cm in the same time.

It should not be thought, however, that flocculation is entirely a phenomenon of physical chemistry. As we have seen in Chapter 2, many organisms contribute to binding together fine sediment particles. Diatoms, for instance, produce extracellular mucoid substances that are particularly important on mudflats, and the production of faecal pellets by worms and snails involves wrapping bundles of fine sediments into packets with a mucoid envelope. The length of these may be measured in millimetres instead of microns, and they may form a layer on the mudflat surface.

Cohesiveness is vitally important in allowing organisms to create and maintain burrows in the sediment that would be impossible in sand. It is also important in determining—and delaying—the degree to which mudflats are eroded. Erosion by water movements occurs quite differently in sand and mud. As water velocity rises over a sandy substratum, the grains begin to move along the bed when a critical velocity is reached: they form a 'bedload' until velocity rises high enough to lift them into suspension. In contrast, a rise in water velocity over a mudflat does not move the particles as bedload. No movement of individual particles occurs, and then when a critical velocity is reached, the sediment breaks up into lumps, and hydrodynamicists speak of 'mass failure'. For this failure to occur, a very high water velocity is needed—so mudflats, once deposited, can only be eroded by higher velocities than are needed to erode sand. Mudflats are, therefore, remarkably stable.

The diversity and distribution of muddy-shore organisms

Muddy sediments harbour an immense diversity of organisms, as well as supporting a very high biomass, unlike sandy beaches. To clarify some of the features of distribution, we consider the organisms in three groups.

Angiosperms and algae

The most obvious organisms, when present, are macroalgae such as the green *Enteromorpha*, whose filaments bind surface mud and sand particles together, and seagrasses such as *Zostera* (eelgrass), whose rhizomes grow beneath the surface. Both occur worldwide. Seagrasses are flowering plants that usually grow low on the shore and stretch into the sublittoral where they may form dense beds. Although *Zostera* is the best known genus, there are many others, including *Thalassia* (turtle grass) in the tropics and *Posidonia* which is widespread in Australia. In the low-salinity waters of estuaries, pondweeds such as *Ruppia* and *Potamogeton* may form abundant beds in the sublittoral, and are often features of brackish pools and ditches.

Somewhat more inconspicuous are the associations of micro-organisms on the surface of the mud, called the 'epipelon'. These consist for the most part of diatoms, cyanobacteria (blue–green algae) and flagellates, and are visible as brown, green or golden-brown films. Many of the genera are cosmopolitan: cyanobacteria such as *Microcoleus* and *Lyngbya*, for instance; diatoms such as *Navicula*, *Nitzschia*, and *Pleurosigma*; and flagellates such as *Euglena*. Little definitive work has been carried out on the distribution of individual species. In general, the diatom communities in different continents contain different species, but some are known to be cosmopolitan. *Hantzschia virgata*, for instance, occurs both on the east coast of North America and in India.

The epifauna

Animals that live on the surface of the sediment (the 'epifauna') cannot always be distinguished from those that live within the sediment (the 'infauna'), but the terms are useful as rough ecological categories. The permanent members of the epifauna are almost all crabs and snails; although many of these burrow at times, they feed primarily on the surface (Fig. 4.1). In warmer areas, *Uca* spp., the fiddler crabs, dominate sandy mudflats around the world. Another genus in the same family (Ocypodidae) is *Macrophthalmus*, which is common in the Southern Hemisphere particularly in South Africa and Australia. In New Zealand, the dominant crab is *Helice*, which belongs to a different family, Grapsidae. All these crabs are active at low tide. On the shores of northern Europe, in contrast, crabs such as *Carcinus* are active only when water covers the flats.

Fig. 4.1 The occurrence of some epifaunal snails and crabs.

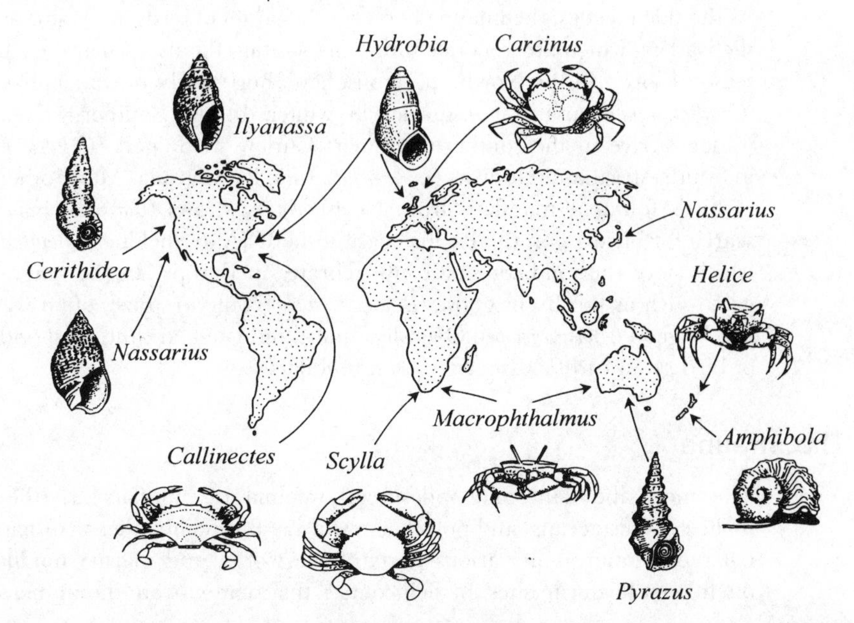

Of the epifaunal snails, a variety of genera dominate in different parts of the world, and they constitute four groups. Perhaps most widespread are the mudsnails—genera such as *Nassarius* and *Ilyanassa*. *Ilyanassa* is native to east-coast USA, but has now been introduced to the west coast. These are whelks, whose relatives are mainly carnivores, but mudsnails cruise over the mud surface eating detritus and also scavenging dead carcasses. Another group consists of purely detritus-eating genera—tall-spired snails such as *Pyrazus* in Australia and *Cerithidea* in western USA. A third group consists of very small snails, also often called mudsnails, which may occur in extraordinary density—particularly *Hydrobia*. These are especially common in northwest Europe, and have given rise to a profuse literature, but related snails occur also in New Zealand, North and South America, and worldwide. Lastly, one of the air-breathing snails, *Amphibola*, is abundant on flats in New Zealand, and has relatives (*Salinator*) in Australia.

Besides the permanent members of the epifauna, large numbers of predatory and grazing animals invade the mudflats at specific times. At high tide, fish such as mullet (*Mugil*) and flatfish such as flounders (*Pleuronectes*) move up the shore. The grazing mullet are found worldwide, but *Pleuronectes* is replaced by other genera in South Africa (*Austroglossus*) and Australia (*Rhombosolea*). Other fish such as eels (*Anguilla*) pass through estuaries in many parts of the world and may be common on mudflats. In addition to fish, crustaceans also invade the flats to feed on the infauna at high tide. Crabs are the most obvious: *Carcinus*, the shore crab, is common in Europe, while *Callinectes*, the blue crab, is found in eastern USA and *Scylla* the mud crab in South Africa (Fig. 4.1). Many species of shrimps such as *Crangon* (in Europe and the USA) and *Penaeus* (in the USA) feed on smaller prey, the meiofauna.

As the tide recedes, the infauna becomes the target of birds. Most important are the waders, which are generally migrants, visiting the flats when their breeding season is over. In the Northern Hemisphere, huge flocks of waders like Dunlin (*Calidris alpina*) arrive in estuaries in winter. In the Southern Hemisphere, waders arrive in the southern summer: Curlew sandpipers (*Calidris ferruginea*) in South Africa and Godwits (*Limosa lapponica*) in Australia. Many other species of birds frequent mudflats, from Northwestern crows (*Corvus caurinus*) which search for bivalves on the Pacific coast of the USA to Shelduck (*Tadorna tadorna*) which sieve the mud for small invertebrates in Europe. In the seagrass beds, geese such as the Brent goose (*Branta bernicla*) come to graze, often with ducks like Wigeon (*Anas penelope*). Overall, waders and geese are the most widespread of bird groups to be associated with mudflats.

The infauna

The most abundant and widespread infauna of mudflats are the bivalve molluscs, crustaceans, and polychaete worms. But many other taxa have infaunal representatives in various parts of the world, and marine mudflats have often been favourite sites for fieldcourses that demonstrate the immense range of invertebrate phyla. Burrowing anemones, e.g. *Edwardsia*, can be common in New Zealand and in east-coast USA, although they are local in Europe. Hemichordates, e.g. *Saccoglossus*, can reach densities of 150 per square metre in west-coast USA. Other worm phyla such as sipunculids, echiurans and nemertines, as well as worm-like burrowing sea cucumbers, are widespread, although not usually found in high densities. In northwest Europe, the brittle-star *Amphiura brachiata* burrows in muddy sand in the littoral and sublittoral, often in high densities; and related species are common in New Zealand and North America. In warmer regions, the brachiopod *Lingula* may occur in dense patches.

Besides the macrofauna, mudflats also harbour abundant meiofauna. Of these the nematodes undoubtedly make up the majority, but harpacticoid copepods may be common, together with turbellarians (flatworms) and two phyla of deposit feeders, the kinorhynchs and the gnathostomulids. Overall, species diversity in mud usually declines drastically as salinity decreases, and while there may be hundreds of species of macrofauna and meiofauna on fully marine mudflats, the number will be reduced to only a few at the heads of estuaries. Few of the macrofaunal taxa mentioned in the last paragraph penetrate far into estuaries, which are reserved for burrowing worms, bivalves, and crustaceans.

Of the crustaceans, the burrowing shrimp *Callianassa* (Fig. 3.9) and its relatives makes its activities most apparent. In South Africa, *Callianassa kraussi* forms craters and miniature volcanoes where it burrows, sifting sediment for food. On the Pacific coast of the USA, *C. californiensis* and *Upogebia* are common in muddy sand and mud, respectively, and *Upogebia* is again frequent at the low-water mark in eastern USA. In New Zealand the dominant genus is *Alpheus*, and in Australia *Trypaea*, but all have a profound effect on the mudflats and their other inhabitants. In Europe, on the other hand, there is no equivalent, and mudflats are often rather flat if they have not been sculptured by lugworms.

Smaller crustaceans, mostly amphipods, do not burrow so deeply and thus do not disturb the sediment so much, but they may be very abundant (Fig. 4.2). *Corophium volutator* in Europe may have densities as high as 100 000 individuals per square metre, and can reach into low salinities. In other parts of the world *Corophium* spp. often inhabit tubes on hard surfaces, but *C. acherusicum* is common on mudflats in South Africa. In eastern USA, *Leptocheirus* constructs permanent tubes in shallow and deep-water mud bottoms, while in western USA, a member of another crustacean group, the tanaids, is common: *Leptochelia* burrows in mud and muddy gravel, and is sometimes found with *Corophium*.

A great range of bivalves can be found on mudflats, from shallow-dwelling species to those that can burrow more than a metre down (Fig. 4.2). Some of the most familiar are the cockles (Cardiidae), which have short siphons and use these for suspension feeding as the shells lie just beneath the mud surface. Examples are *Cerastoderma* in Europe and *Clinocardium* in west-coast USA. In parts of the Southern Hemisphere, their place is taken by species of *Anadara*, which belong to a quite separate family (Arcidae). These have no siphons at all,

Fig. 4.2 The occurrence of some infaunal bivalve molluscs and crustaceans.

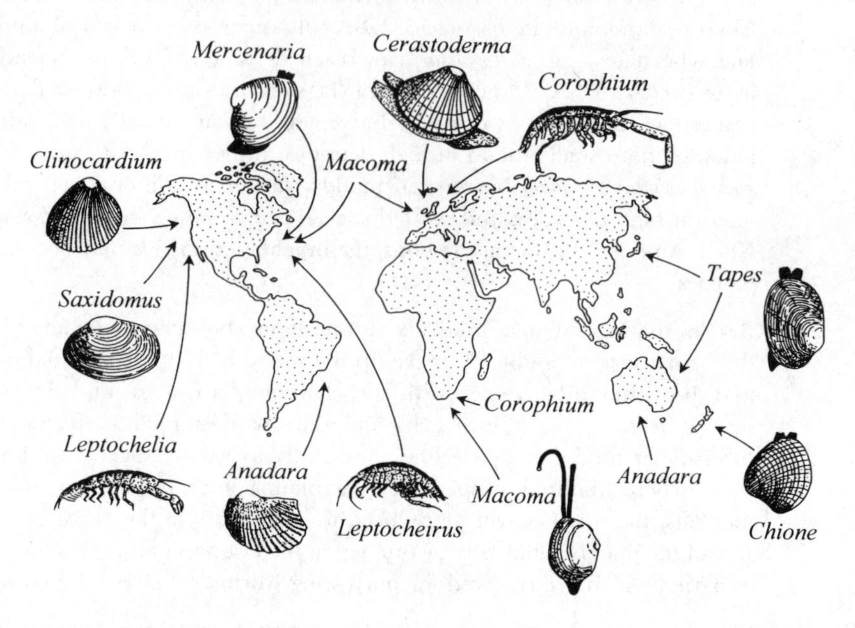

but draw water in through the mantle edges and also suspension feed. Many other suspension feeders around the world belong to the Veneridae, which again have short siphons and lie just beneath the mud surface: *Chione* in New Zealand, *Mercenaria* in east-coast USA, *Saxidomus* in west-coast USA, and *Tapes* in Australia and Japan.

Lying slightly deeper in the mud are the deposit feeders with long mobile siphons which can reach to the surface. Most of these forms are in the family Tellinidae. *Tellina*, for instance, can be very common, and *Macoma* is abundant even in the reduced salinities of estuaries.

The third category of bivalves consists of deeper burrowers. Some of these have extraordinarily long siphons which reach to the surface to allow suspension feeding, but can be retracted upon disturbance. *Mya*, for instance, is found in Europe and in east-coast USA and *Panopea* in west-coast USA. Other species form burrows and can move up and down in them very rapidly. Some of these such as the razor shells, *Ensis* and *Solen*, are suspension feeders. Others, like *Solemya*, common in Australia, New Zealand, and east and west coasts of North America, form U-shaped burrows in anoxic sediments. They have reduced guts and depend upon symbiotic bacteria to provide energy (see p. 71).

Polychaete worms are usually abundant burrowers in marine and estuarine mud. Some are active predators—the so-called 'errant' polychaetes—and of these several species are cosmopolitan. *Scoloplos armiger*, for instance, lives in Europe, the Pacific, the Arctic, and the Antarctic. Many species belong to the families Nereidae or Nephtyidae. In the Nereidae, for example, is the large cosmopolitan *Neanthes virens*, and the well-known estuarine species *Nereis* (*Hediste*) *diversicolor*, which is really an omnivore. Another cosmopolitan genus is *Glycera*.

The more sedentary, deposit feeding polychaetes include a wide variety of families. Some species are cosmopolitan, such as *Capitella capitata* (family Capitellidae), often an indicator of organic pollution. Other common families are Maldanidae or bamboo worms, containing *Clymenella*, common in eastern USA; the Cirratulidae, distinguished by their long thread-like red gills; the Terebellidae, fat worms with many long thin tentacles that spread out on the mud surface to feed; and the Arenicolidae or lugworms, often used for bait around the world. *Arenicola* spp. are major 'bioturbators', turning over and disrupting the structure of the mud, and so are important in affecting many other infauna.

Organisms and their adaptations

To give some background for discussions of the ecological processes that occur on mudflats, we now discuss briefly some of the ways in which organisms have become adapted to life in fine sediments.

Diatoms and flagellates: how to stay in one place by migrating vertically

During daytime low-tide periods, marine and estuarine muds often turn a rich green or brown colour. The colour is due to diatoms and photosynthetic flagellates such as *Euglena*, which migrate upwards from depths of 1 to 2 mm to lie on the sediment surface. The organisms remain here until about an hour *before* the tide begins to cover the flat, then burrow again, and this 'forecasting' of the tidal time suggests that movement is determined by an intrinsic rhythm. The existence of a rhythm is confirmed by bringing diatoms and euglenids back to the laboratory and examining them in conditions of constant light and under no tidal influence. Here they continue to show a migratory pattern, emerging at times when it is daylight outside (Fig. 4.3). As we shall see, however, this diurnal rhythm is not sufficient to explain all their movements, because tidal variation does not coincide with the day–night cycle.

Most observations of diatom migration have been made by placing lens tissue on the sediment surface and using this as a kind of trap: the diatoms move into it and can then be conveniently removed. This technique has its limitations, however, and with the use of low-temperature scanning electron microscopy, diatoms can be viewed *in situ* on the sediment surface and below it (Paterson, 1986). Species such as *Scoliopleura tumida* can be seen to form layers two to three cells thick, completely covering the mud surface at low tide. When the tide rises, a few individuals remain on the surface (undetected by the lens tissue technique), suggesting that there is a random component to movement as well as directed migration towards the light.

Fig. 4.3 Vertical migration by photosynthetic mudflat organisms. Observations on the flagellate *Euglena obtusa* and the diatom *Cylindrotheca signata* were made in the laboratory in constant light. Black portions of the bar at the top of the diagram show times of natural darkness. Cell counts show numbers of organisms found in squares of lens tissue (0.25 cm^2) placed on the sediment surface. Both organisms migrate upwards at times of natural light. (After Round and Palmer, 1966.)

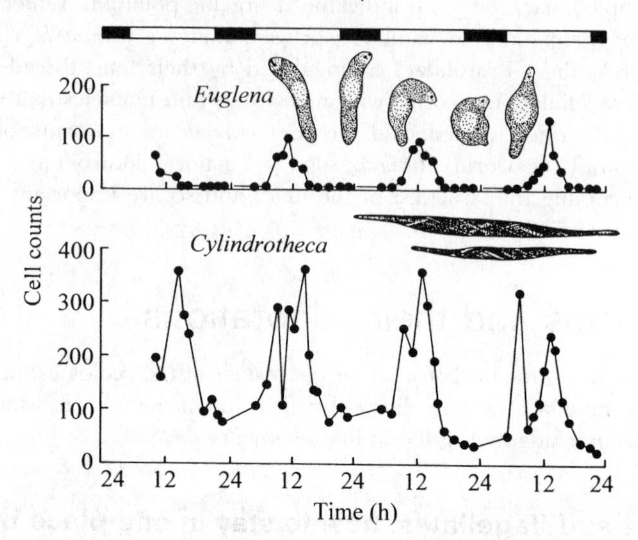

Estuarine diatoms are not alone in showing a pattern of vertical migration—diatoms in freshwater muds also move to the surface in the daytime. But in tidal environments the movements have to be geared to times of low tide as well as times of daylight—and the tides advance by 0.8 h each day (p. 9). How do migratory rhythms cope with this problem? The activity of most diatoms and euglenids is under the control of a diurnal rhythm which is itself kept in check by light intensity. When muddy estuarine water rises over the flats in the natural environment, light is prevented from reaching the mud, and the rhythm is suppressed—so diatoms come to the surface only when the tide is out. This hypothesis does not explain why diatoms move downwards *before* the tide reaches them, but at least one species of diatom has a tidal rhythm as well as a diurnal one. *Hantzschia virgata*, from east-coast USA, displays a rhythm in the laboratory that moves forward each day in synchrony with the tides, even under conditions of constant light and no direct tidal influence (Round, 1981).

Diatoms move rather slowly by extruding mucilage which 'pushes' them along. Why should they expend energy on migrating distances which may be more than 20 times their body length? One half of the answer to this question is that as the tide rises, they may not be able to maintain their position, even with the aid of their secreted polysaccharide 'glue' (see p. 24). So those that do not burrow are swept away in the rising water. The second half of the question concerns the light regime in the top mm of mud. Can the organisms receive enough light for photosynthesis if they remain below the surface? Measurements in fine silts show that at 0.4 mm down, light may be reduced to 1% of its intensity at the surface (Jørgensen and Des Marais, 1986). At this intensity, diatoms could survive, but not grow—so they *must* migrate upwards to receive sufficient light.

Seagrasses: underwater meadows

The majority of flowering plants live on land, but in some areas seagrasses dominate the low-tidal levels of sedimentary shores and may spread into the sublittoral. They can colonize both sand and mud by sending out rhizomes—tuberous structures that can produce new bundles of shoots and leaves, and also store food supplies. Although new shoots may later become disconnected from the original plant, they are genetically identical to it, so large areas may be covered by plants with the same genetic make-up. These constitute what is called a 'genet'. Although individual offshoots may die, the genet may continue indefinitely.

Seagrass plants are very productive. Some tropical species have average production rates of 800 g dry weight/m^2 per year, and the leaves of *Thalassia*—turtle grass—can grow 2 cm in a day. The leaves grow from the base, while the tips erode to form detritus, and most seagrass biomass enters the food web in this form rather than being consumed directly as living material: maybe only 10% of seagrasses are eaten before they decay.

Nevertheless, some *Zostera* species are immensely important for specific herbivores. In east-coast USA, the American Wigeon, *Anas americana*, feeds over sublittoral seagrass beds, dabbling from the surface; and in Europe *Anas penelope* does the same as well as grazing on plants left out of water at low tide. Diving ducks such as the Canvasback duck of east-coast USA also feed on *Zostera*, and some species of geese are practically dependent upon it during winter. Such dependence has caused serious declines in bird populations following declines in *Zostera*. In the 1930s, a fungal disease spread through the *Zostera* beds of Europe and east-coast USA, and many of them disappeared. As a consequence, the Brent goose in Europe declined by 75%, while populations in the USA fell by 90%. Although the *Zostera* began to recover in the 1950s, it was damaged by the severe winter of 1962/3, and has not regrown to anything like its former extent. Some birds have now switched to feeding inland on grasses, and both Wigeon and Brent geese now feed to a great extent on permanent pasture.

Seagrasses are important components of soft-shore communities far in excess of their contribution to the diet of herbivores. This importance becomes apparent after even a brief examination of a seagrass bed: the fauna here is both more diverse and more abundant than that in nearby unvegetated sediments. Why should this be so? For a start, seagrasses offer both a greatly increased surface area, and a relatively firm substratum in comparison with a bare mud surface. The leaves of *Zostera* become encrusted with sessile epiphytes ranging from algae to bryozoans, and a whole mobile community of nemertines, snails, amphipods, etc., develops. Between the leaves and shoots, seagrasses also offer a sheltered habitat. Here a community of larger mobile animals can live, including crabs, shrimps, and fish. Experiments in Australian seagrass beds have shown that these mobile animals prefer the most dense beds, but whether this choice is governed by the protection the beds give against predators, or by some other factor, is unclear. The physical structure of the beds appears to be particularly important to fish, as these will move into artificial beds as well as real ones. But fish do not necessarily inhabit seagrass beds after they have become adults. The King George whiting, a commercial fish species in southern Australia, uses seagrass beds as nursery areas, but moves out on to clean sand after about 4 months (Fig. 4.4). Besides shelter, seagrass beds offer a rich source of detrital food, especially because the majority of seagrass detritus usually remains within the bed. Large communities of infaunal detritivores take advantage of this supply, and the mud between the rhizomes abounds in polychaetes, bivalves, snails, and amphipods.

Seagrass meadows are vulnerable to attack, and in east-coast USA sharks and rays can denude large swathes as they dig up the bottom searching for clams. Such denuded areas are usually recolonized quickly, but long-term die-back causes more concern, because the beds stabilize sediments as well as acting as nurseries for fish. Unfortunately, the causes of die-back are far from clear (Underwood and Chapman, 1995). Sewage effluent may cause excessive growth of epiphytic algae on the seagrass leaves; dredging may alter the local wave climate and thus may lead to erosion of the beds; industrial effluents and oil spills may have direct toxic effects. But the relative importance between these factors has yet to be worked out, and other influences may still await detection.

Fig. 4.4 Abundance of the King George Whiting, *Sillaginodes punctata*, at one site in Port Phillip Bay, southern Australia. Newly settled individuals are common on algal-covered reefs (open circles) and seagrass beds (filled circles). Older juveniles move on to unvegetated sands (triangles). Bars show S.D. Data from Jenkins and Wheatley (1998).

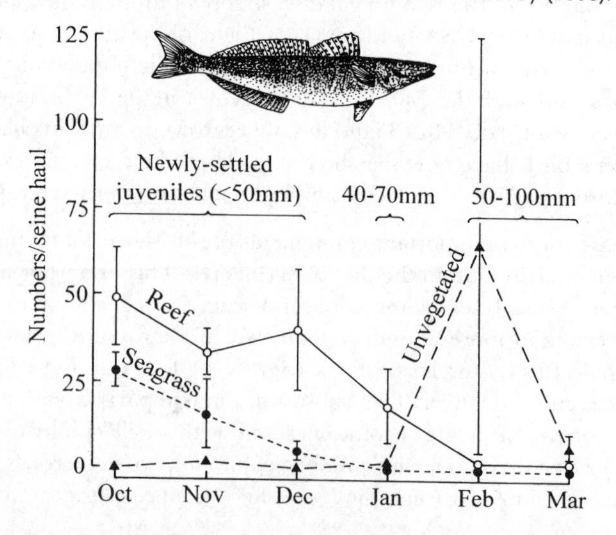

When to be active

Suspension feeders must by definition be active when covered by the tide. But for deposit feeders great variation is found in the timing of activity periods. On European mudflats, there is usually little activity after the mud surface dries out at low tide. But in America, Asia, and Australasia the mudflats become scenes of activity when the tide is out. *Uca* spp. often close their burrows to prevent the entry of water at high tide, and emerge to feed when the mud surface is damp but not covered by water. In New Zealand, the grapsid crab *Helice crassa* exhibits a more complex picture (Fig. 4.5). Direct observation shows that *Helice* is active, feeding near to its burrows and excavating, at low tide. When pitfall traps are used to study its activity, though, the major catches are made at times of *high* tide. This is the time when the crabs show exploratory behaviour, and hence are caught in the traps. When *Helice* is examined in the laboratory, most movement is in fact shown at times of high tide—and this pattern is maintained in continuous dim red light and with no tidal cues, so is evidently under the control of a tidal rhythm.

Lower down on the New Zealand flats, the pulmonate snail *Amphibola crenata* is active at times of low tide, crawling and feeding until the mud dries out. It then

Fig. 4.5 Activity patterns of the New Zealand crab *Helice crassa*. The upper graph shows catches in pitfall traps on the shore: activity occurs both at LW (low water) and HW (high water—filled arrow). The lower graph shows activity of 10 crabs in the laboratory, in continuous dim red light and with no tidal rise or fall: activity is primarily at HW. Open arrows show the times of high water the crabs would have experienced on the shore. Dotted lines show when no observations were made. Horizontal bars show light (white) and dark (black). (After Williams *et al.*, 1985.)

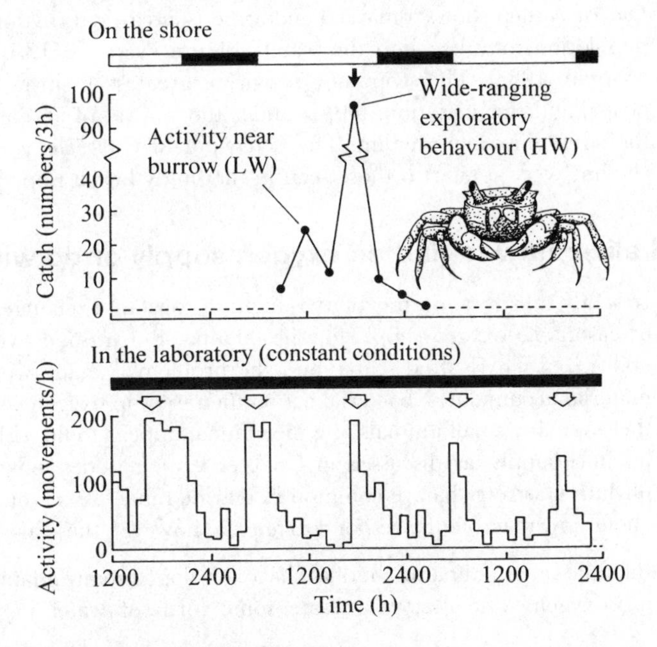

burrows, and remains burrowed during the period of high tide. This is quite a different pattern from that shown by many other snails, and probably relates to the fact that *Amphibola* breathes air (through a lung), while most snails take oxygen from water (through a gill). The European *Hydrobia ulvae*, for instance, is a gill breather. It is abundant on estuarine mudflats, crawling and feeding under water and when the mud surface is damp, climbing plants in search of food, but burrowing or withdrawing into its shell when the flats dry out at low tide. Occasionally, the returning tide may displace snails that have climbed upwards, so that they float on the surface water film in calm conditions, but this has little significance for the population in terms of dispersal. *Hydrobia* shows no evidence of any endogenous tidal rhythm, and only a faint circadian (24.5 h) rhythm. Instead it appears to react directly to environmental stimuli. It shows maximum activity when covered by water, during darkness (Barnes, 1986). This picture contrasts strongly with that on rocky shores, where most snails have endogenous rhythms, but we do not know enough about other mudsnails to be able to generalize about rocky shore/muddy shore differences.

Overall it is likely that most muddy-shore animals have strong endogenous rhythms which govern their behaviour, rather than reacting directly to external stimuli. One of the crustaceans often found with *Hydrobia* is the amphipod *Corophium volutator*, and this shows a great dependence upon internal rhythms. *Corophium* spends most of its time in a burrow, but adults, particularly males, crawl over the mud surface and sometimes swim on the ebbing tide, in search of mates. As neap tides progress to springs, juveniles also swim, usually on night-time high waters, and this activity ensures dispersal of the population (Lawrie and Raffaelli, 1998). If the pattern of swimming is examined in the laboratory, it becomes more obvious when animals are deprived of their muddy substratum. *Corophium* then shows a marked endogenous circa-tidal rhythm, and 50% of the population may rise into the water column every 12–13 h (Holmström and Morgan, 1983). The swimming activity is greatest at spring tides and least at neaps, and this variation, too, is under the control of an endogenous rhythm, the circa-semilunar rhythm. The behaviour of *Corophium* is thus controlled by rhythms very similar to those seen in the sandy beach isopod *Eurydice* (p. 42).

Buried alive: how to find an oxygen supply or do without it

In sandy substrata, permeability is high, so constant exchange of water provides a reasonable oxygen supply for the infauna. But in mud, with its low permeability, oxygen is often scarce and the problem is made worse when organic material accumulates: bacterial action then uses up oxygen very quickly. Nevertheless, many small animals, the meiofauna, appear to flourish where oxygen is in short supply, as discussed in Chapter 2: nematodes, oligochaetes, and the phylum Gastrotricha are common in marine muds. As so much more is known about how macrofauna respond to anoxia, however, these are discussed instead.

Infaunal macrofauna on mudflats have developed many adaptations for obtaining oxygen. The majority create some form of water current, drawing in aerated water when the tide is high, then storing the oxygen to carry them over the low-tide period, meanwhile reducing their activity to ensure they use as little oxygen as possible (Newell, 1979). The so-called 'sedentary' polychaetes show a wide variety of ways of respiring. The fan-worms, such as *Sabella*, have a crown of tentacles that protrudes into the water and acts both as a gill and as a suspension feeding net. This is supplemented by peristaltic pumping of water through the tube in which the worm lives. The lugworms like *Arenicola*, on the other hand, depend upon drawing water into the burrow and so irrigate their gills without ever emerging above the sediment surface. Similarly, *Clymenella* can live in very anoxic black sediments by producing a water flow through the burrow: it moves up and down, drawing water in and out.

None of these worms have any specific mechanism for obtaining oxygen when the mudflats dry out, but some of the bivalves have adaptations that allow them to breathe air. When the cockle *Cerastoderma* (*Cardium*) *edule* is left dry by the tide, it claps its two shell valves together to expel water from the mantle cavity, then draws in air and absorbs oxygen. By doing this it can maintain its rate of respiration equal to the lowest rate found when it is under water (Boyden, 1972).

Most infauna depend upon the water for their oxygen supply. But what happens if the water itself is low in oxygen? The infauna varies in its responses to such stress: *Arenicola*, for example, is resistant to severe hypoxia while other polychaetes such as *Capitella* can resist only mild hypoxia and many crustaceans such as the epibenthic prawn *Crangon* are extremely sensitive. Many enclosed marine and estuarine areas are now showing degrees of lowered oxygen, or hypoxia, mainly caused indirectly by human intervention (see Chapter 11). Severe hypoxia, such as that seen in the Chesapeake Bay and parts of the Baltic Sea, has led to mass mortality of the benthos (Diaz and Rosenberg, 1995).

Perhaps more difficult to withstand than the direct effects of oxygen lack are the rising levels of sulphide that occur when sediments become anoxic. When marine sediments run out of oxygen, sulphates are reduced to sulphides, and hydrogen sulphide in particular is very toxic. How do animals cope with the situation? The burrowing shrimp, *Callianassa subterranea*, lives sublittorally off the west coast of Scotland, UK, in muddy sediments, and the water in its burrows may accumulate up to 200 mM sulphide. *Callianassa* seems to survive by oxidizing the sulphide that is taken into the body into thiosulphate, which is much less toxic. This detoxification mechanism in fact allows the shrimps to tolerate as much as 1 mM sulphide for up to 24 h—a remarkable ability as most decapod crustaceans are extremely sensitive both to oxygen lack and sulphide build-up.

Even more extreme examples are found in the bivalve molluscs. Two families have actually turned the problem of sulphide detoxification into a benefit. When the incoming sulphide is oxidized in their gills, the process generates energy, and this energy can be used by symbiotic bacteria that live in the gills. The bacteria use the energy to fix carbon dioxide, and the resulting carbohydrates provide the host bivalves with a food supply. The bivalves are thus extremely tolerant of high sulphide levels and live in rather extreme conditions.

Lucina, for instance, lives in anoxic muds below the tropical seagrass *Thalassia*. *Solemya reidi* is found in anoxic silts on the Pacific coast of North America. It has even been collected from such insalubrious habitats as deposits of human hair in the Los Angeles sewer outfall! Here it inhabits U- or Y-shaped burrows which it periodically ventilates to obtain oxygen (Fig. 4.6). Then it lies quiescent while sulphide diffuses into the burrow from the surrounding sediments. This species depends entirely on its symbiotic bacteria for food, and has no gut (Reid, 1998).

How to grow and multiply: reproductive strategies

There is an extremely wide range of lifespans in mudflat animals, and a variety of reproductive strategies. Some species are annuals, while others may live for a quarter of a century or more. Some breed once and then die—the so-called semelparous species. Others breed many times—the iteroparous species. Some have pelagic larvae, while others brood their young which hatch as miniature versions of the adults. What governs all this variation on the mudflat? Many features of reproduction are determined by overall phylogenetic patterns: in general, for instance, amphipods brood their young, while in many polychaetes the eggs are released from the females by rupture of the body wall, and the adults then die. But many other features, including the growth rate and the amount of parental investment in reproduction appear to be related more to environmental factors, both physical and biological.

The amphipod *Corophium volutator* seldom lives for more than 1 year, and often only for 6 months. The females carry the young in a brood pouch until they are ready to be released. In one season a female may produce up to five broods, and as the young are liberated they form small side burrows off the parental one until they are large enough to inhabit their own. Other species of *Corophium* have similar life histories.

Fig. 4.6 The bivalve *Solemya reidi* in its burrow, when ventilating and not ventilating. It obtains sulphide (HS^-) from anoxic sediments when not ventilating, and uses oxygen when ventilating to oxidize this to sulphate. RPD shows the redox potential discontinuity. DOM, dissolved organic matter. (After Reid, 1998.)

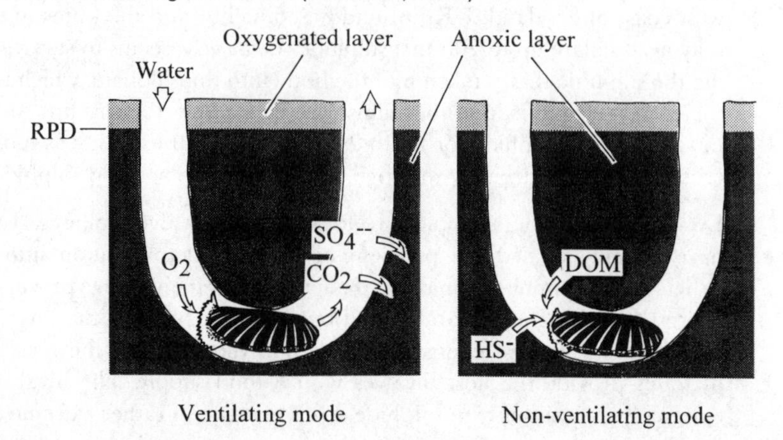

In contrast, the nereid polychaetes show a diversity of reproductive habits. *Neanthis succinea*, the common clamworm of the Chesapeake Bay, breeds like many other marine polychaetes. It metamorphoses into a swimming stage, the heteronereis, in the breeding season, and these stages swim to the surface to release their gametes, then die. The larvae spend some time in the plankton before settling on the mud again. The common European estuarine polychaete *Nereis diversicolor*, which lives for about 2 years, also dies once it has spawned. The larvae of *Nereis* can swim, but are said not to enter the plankton but to remain near the substratum—presumably an adaptation to life in estuaries, where pelagic larvae could easily be swept out to sea, to unfavourable habitats. In Australia, another estuarine species, *Ceratonereis limnetica*, spawns on the estuary bed and then broods the eggs in tubes within the sediment. This procedure is presumably another adaptation to life in areas of fast currents where larvae could be swept away.

Some of the molluscs exhibit much longer lifespans, and the bivalves in particular have been studied because it is often possible to tell an individual's age from the number of 'growth-check' lines on the shell, formed each winter. *Macoma balthica*, for instance, which is widespread from northwest Europe to the west coast of the USA, can be aged fairly easily. In the Baltic Sea its age varies greatly. Shallow-water individuals may live for 7–8 years, and grow rapidly to a length of up to 16 mm. At depths of 35 m, however, some individuals may live for 30 years, but they grow only slowly to about 13 mm. The difference in growth rate is probably caused by a temperature effect: at depth in the Baltic, the mean annual temperature is only 4°C, and for much of the time it is near zero. But the difference in ultimate size is unlikely to be due to temperature: as a general rule, marine invertebrates are larger in colder waters. *Macoma* breeds annually (i.e. is iteroparous), but surprisingly recruitment to the deep populations may fail for periods as long as 25 years (Segerstråle, 1962). Why should this happen? The interaction between *Macoma* and the amphipod *Pontoporeia* has already been described (p. 23). It seems that at deeper levels in the Baltic *Pontoporeia* may effectively consume all the settling larvae of *Macoma*, preventing recruitment. Only when *Pontoporeia* populations fall can the bivalve population obtain a foothold.

Small mudsnails of the family Hydrobiidae are often common on mudflats and in coastal lagoons. In northwest Europe, four species are common, but a comparison between two is particularly instructive. *Hydrobia ulvae* lives primarily on tidal estuarine mudflats, but is sometimes found in non-tidal coastal lagoons, while *Hydrobia ventrosa* occurs only in lagoons. Both species are essentially short-lived snails that have only one main breeding period, i.e. they are semelparous. Both live for 2 years or a little longer, breeding in their second year. *H. ulvae* produces egg capsules which contain up to 40 small eggs, and these hatch into larvae, the veligers. The veligers spend some time in the plankton before settling and metamorphosing into snails. *H. ventrosa* produces egg capsules with a few relatively large eggs, which hatch directly into young snails. These different strategies presumably fit the snails for their respective habitats. For *H. ventrosa*, in landlocked lagoons, pelagic larvae would not help dispersal; while for *H. ulvae* on tidal mudflats, a proportion of the larvae that spend up to 3 weeks in the plankton may well be swept away, but others can found new populations.

How much investment do the snails make in reproduction? Here an interesting comparison can be made between populations of *H. ulvae* on mudflats and in lagoons (Fig. 4.7). The total weight of eggs produced over the year by females in mudflat populations is enormous—something like 1000% of the body weight. In lagoonal populations, females produce only the equivalent of 50%, and this figure is very similar to that produced by lagoonal *H. ventrosa*. Why this enormous difference? The answer may lie in the different environmental pressures acting in open tidal flats and in closed tideless lagoons (Barnes, 1994*a*). On the flats, predation by wading birds can be intensive, especially in winter (see p. 85), while in summer the snail populations can suffer severe desiccation because they can be out of water for 70% of the time. Adults may therefore not live long, and must put as much energy into reproduction as possible if they are to pass on their genes to the next generation. In lagoons, the situation is different. Because the populations are always submerged, predation by waders is non-existent, there are few fish predators to take their place, and there is no desiccation problem. Adults can therefore put more energy into growth, which may aid long-term survival.

Winter visitors: wading birds as tourists

The invertebrate fauna of mudflats provides a substantial food source for birds of the suborder Charadrii—waders as they are called in Europe, or shorebirds in America. Over the period of the Northern Hemisphere winter, mudflats in both hemispheres become populated by large numbers of birds such as plovers,

Fig. 4.7 Cumulative weight of eggs produced by two species of mudsnail in eastern England. Intertidal populations of *Hydrobia ulvae* (filled circles) invest far more in eggs than do lagoonal populations of the same species (open circles), or of *H. ventrosa* (triangles). (After Barnes, 1994*a*.)

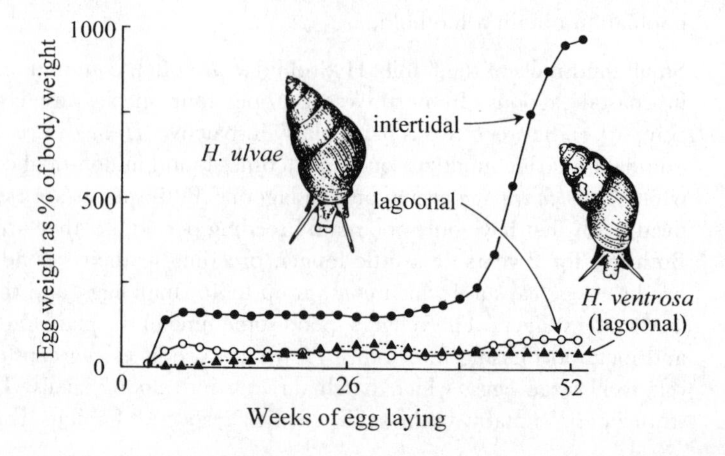

dunlin, knot, godwits, sandpipers, and many others. The actual timing of their arrival and departure varies for species and for different areas, but waders are usually abundant from something like September to April on estuarine flats in Brazil and Australia as well as in Europe and North America. Why do these tourists stay only for the period of the northern winter?

Few waders breed in their wintering estuaries. Some migrate locally to inland moors or bogs, but many fly enormous distances in the spring to breed in the Arctic. They use well-defined 'flyways', including the east and west coasts of America, and the west coasts of Africa and Europe (Fig. 4.8). Some individuals of the Knot, *Calidris canutus*, for instance, fly from Argentina to the Canadian arctic, while those wintering in southern Africa and Europe fly north to Siberia or Greenland. The trip may be as much as 15 000 km and is usually accomplished in a series of long 'hops'. The birds stop at a number of staging posts where they can feed and re-fuel for the next flight.

Feeding on estuarine flats, both those of the wintering grounds and the staging posts, is thus extremely important and protection for these sites is essential if the birds are to be conserved (see Chapter 11). Knot can put on weight on the feeding grounds at rates as high as 3 g/day, mostly in the form of fat but partly as protein. Individuals start at a total weight of 140 g before the build-up for migration, but may weigh 220 g by the time they fly. Waders may therefore have significant effects on mudflat invertebrates over the winter period (p. 89).

If they can feed so well in estuaries and flats far to the south of their breeding grounds, why do they not breed there? Why migrate at all? There are no simple answers to these questions. Waders probably breed in the arctic at least partly for historical reasons. Cold-adapted species evolved during the Pliocene, 1.5 million years ago, as cold regions moved southwards over the globe. Then

Fig. 4.8 Major flyways of waders and wildfowl from their arctic breeding grounds to wintering grounds in the south (open arrows). Filled arrows show routes of Knot (*Calidris canutus*) from Ellesmere Island, Greenland and Siberia to southern Africa. Note that separate populations of Knot also winter in America. (After Ferns, 1992.)

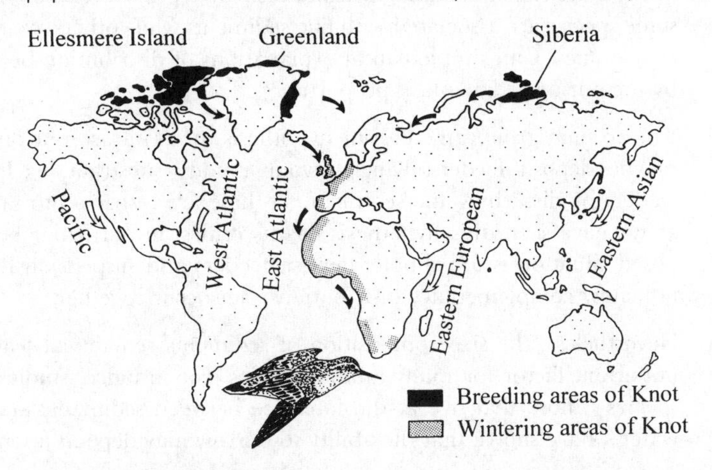

during the Pleistocene many populations were separated when their habitats contracted during inter-glacials (Hale, 1980). Although it has subsequently been possible for them to retain their arctic breeding grounds, it is impossible for them to live there for the rest of the year because they are frozen over—hence the migration south. Life in the south provides rich pickings for fully fledged birds. But it is possible that as breeding grounds the southern estuaries are not ideal. Predators abound, and it may be that chick survival would be very low (Evans in Jones and Wolff, 1981). In contrast, predators are few in the arctic, and this may explain why waders expend massive amounts of energy migrating extraordinary distances.

Distribution on the mudflats: Why are organisms found where they are?

The problem of what governs distribution on the shore has been discussed by marine biologists over a long period, and it has probably engendered more argument than any other. We have seen that animals on sandy shores occur in vertically defined zones (Chapter 3), and the same can be said of muddy shores, but the factors determining such zonation are still far from clear. Physicochemical features such as the period of water coverage, salinity, drainage, and organic content of the sediment are important, but they act together with biological features such as competition and predation to produce a complex picture. Animals on muddy shores also show a patchy or mosaic distribution, like those on sand and rock, so even the measurement of distribution patterns is not simple: different scales of recording will produce different pictures. Two approaches that have been used are now discussed.

Particle-size composition

Early work on mudflats showed correlations between particle size composition of the sediments and faunal distributions. On a gross level there is no doubt that some species are associated with fine sediments while others are found mainly in coarse ones. Can simple causal explanations of distribution be derived merely by measuring sedimentary properties?

The trophic group amensalism hypothesis (p. 24) arose out of this approach: maybe deposit feeders, which favour muddy substrata, exclude suspension feeders by disturbing the sediment, so these are restricted to sandy areas? But as we have seen, this hypothesis has its critics, in particular because the proposed situation is too simple: deposit feeders and suspension feeders often do not show complementary distributions, but occur together.

Nevertheless, the size composition of sediments remains at least a potentially important factor for many animal species. For instance, studies on epibenthic species—those that live at the interface between sediments and the overlying water—have shown that the ability to burrow may depend upon how coarse or fine the sediment is. When the European shrimp, *Crangon crangon*, is placed on various grades of sediment in the laboratory, it is most successful at burying itself in medium grades. It uses its appendages to scrape a furrow in the sediment, then flicks sand over the carapace until only the antennae and eyes are showing. In very coarse grades it does not burrow, and in fine sediments it only partly burrows (Fig. 4.9). Other species have a more 'generalist' approach—the plaice, *Pleuronectes platessa*, can bury itself over a much wider range of sediment types.

Similar effects may be found in the laboratory when animals are given a choice of sediments. When the mudsnail *Hydrobia ulvae* is offered a selection of natural substrata from fine mud (median diameter <38 μm) to muddy sands (median diameter approximately 250 μm), it is attracted mostly to the fine mud. Further tests show that *Hydrobia* is very sensitive to sediment composition—but in the field its distribution is not correlated with particle size (Barnes and Greenwood, 1978). Evidently other factors overrule this choice. In Chapter 2 we emphasized that particle size is not the only important property of sediments. The degree of shelter and 'stability', the boundary-layer flow and populations of micro-organisms all affect the benthos (Snelgrove and Butman, 1994). Desiccation, salinity, availability of detrital food, and many other physical variables may be involved. In addition, as we have discussed for sandy shores, biological interactions such as competition, predation, and parasitism may be important, and we examine them later (p. 83). In the meantime, we consider a factor that we have not yet discussed: the relative importance of the adults, juveniles, and larval stages in governing patterns of distribution.

Who chooses—adults, juveniles, or larvae?

Many studies of invertebrate distribution on mudflats have centred upon the adults. In the case of mobile fauna, adult choice is probably very important—but many members of the benthos produce pelagic larvae which settle on to the

Fig. 4.9 The effect of particle size on the burying ability of a 'substratum specialist', the shrimp *Crangon crangon*, and a 'substratum generalist', the plaice *Pleuronectes platessa*. Data are for animals of similar sizes (55–60 mm). (After Pinn and Ansell, 1993.)

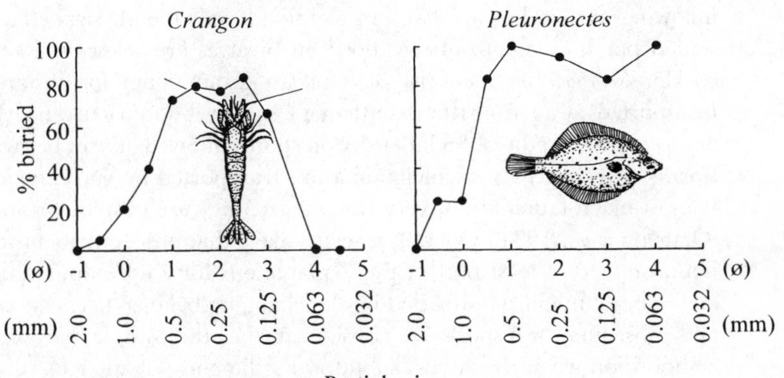

mud surface and them metamorphose into juveniles. If the adults are sessile, choices made by the larvae will determine where recruitment to the population occurs. In more mobile species, it may be the juveniles who make the choice: post-settlement stages are often able to move around until they reach an appropriate habitat. Even if they cannot move significant distances, differential mortality of juveniles at different sites may determine population distributions.

It is often not at all clear just how much choice larvae can make. To some extent, they may be passively distributed by currents (see Chapter 2). Experimental observations in still water have often suggested that larvae show some habitat specificity; but still water seldom occurs in the sea, and to be realistic observations are probably better made in laboratory flumes, where current speed can be adjusted. Under these conditions, some species like the polychaete *Capitella* can select particular sediments even when constrained by currents (Snelgrove and Butman, 1994), so it seems likely that larvae are able to have at least some influence on distribution. Because of the difficulties of distinguishing in the field the effects brought about by larval choice and differential larval mortality, however, recent debate has focused at a slightly coarser level. If we could decide whether soft-sediment communities are structured by availability of larvae (so-called 'recruitment limitation'), or by post-settlement processes such as competition, predation, and physical disturbance, we would at least narrow down the options.

The processes that lead to the supply of a 'rain' of larvae on the sea floor are often referred to under the title of 'supply-side' ecology. What do we know about how this governs benthic populations? A good example is provided by the bivalve *Macoma balthica*. Like other bivalves, this shows some years of intense larval settlement and others in which very few larvae arrive on the mudflats (p. 73). While the high-settlement years often result in an abundant year-class which can be followed through succeeding years, a strong year-class can also be formed when there are relatively low levels of settlement (Olafsson *et al.*, 1994). In this example at least, the relationship between supply of larvae and the adult population is not generally a good one.

Can we turn to post-settlement processes for an explanation of how populations are limited in abundance? We deal with some biological interactions in the following section, but here we can mention briefly the physical effects on newly settled benthos. Again, observations on bivalves are relevant. Recruits of the cockle, *Cerastoderma* (*Cardium*) *edule* up to 2 mm long, for instance, can be transported away from their settlement site by bottom currents. Juveniles of *Macoma*, which settle at high density on shores in Sweden, can be swept away in storms. Even permanent meiofauna are transported by currents, and as juveniles of macrofauna are of very similar size, they are probably transported too (Olafsson *et al.*, 1994). Overall, it seems likely that post-settlement movement is important, so at least part of the explanations for adult population density at any one site might be provided by the changing balance between larval supply and subsequent transport. We consider further the complexities of community composition when we discuss sublittoral sediments (Chapter 6).

Food supply and biological interactions

One of the major sources of food in mudflat areas is detritus. We begin by considering what detritus is, then discuss its consumers and higher trophic levels.

Detritus

Detritus has been defined as the organic carbon lost from any trophic level, excluding losses by predation. It includes plant material that dies and decays, soluble compounds exuded by plants, animal excretions and secretions, as well as dead animal remains. In particular, dead plant material from land run-off and from salt marshes contributes to the detrital pool because little macrophyte material in the marine benthos is eaten directly while alive—in contrast to the situation on land where insect herbivores consume angiosperms voraciously. In the spring, particulate detritus rains down from dying phytoplankton blooms in the surface waters of the sea, and these blooms also release soluble organic products. Detritus thus contains both solid material—the particulate organic matter (POM)—and dissolved organic matter (DOM).

One of the major problems in understanding how detritus is utilized by detritivores is that it is continually changing: bacteria, protozoans, and fungi attack organic material as soon as it is released, altering its composition, including such features as its C/N ratio (see p. 19). Microbial populations build up within the detritus, and these populations are often enhanced by the detritivores themselves: as detritivores feed, they tend to break up detrital particles into ever smaller size fractions, and thus they expand the surface area available for microbial attack. As particle size decreases, microbial biomass increases (Fig. 4.10). This process

Fig. 4.10 Microbial density in relation to particle size of detritus. The detritus originated from the sea grass *Thalassia*. Inset shows the microbial community on a detrital particle. Ciliates (triangles), flagellates (filled circles), and bacteria (open circles) all have highest densities on small particles. (After Fenchel, 1970.)

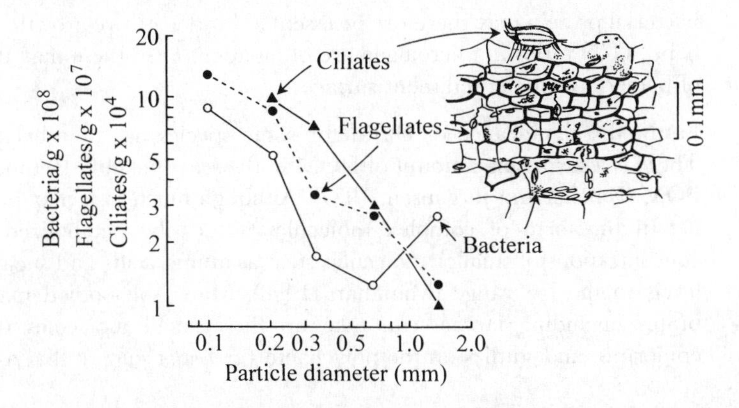

of bacterial culturing can be thought of as a form of 'gardening', akin to the procedures utilized by the lugworm (p. 20).

Detritus cannot therefore be viewed as a simple azoic mass of organic material derived from dead remains. It is a constantly changing mixture of dead material and a living culture of microbes. This complication has led to intense discussion about how detritivores gain their energy supply. Do they eat the dead organic material, or do they digest the microbial fraction?

Views on this question have fluctuated over several decades. Initially, dead POM and DOM were seen to be the major energy sources. Then bacteria were suggested as the most important detrital fraction. At the present time, the situation is seen to be slightly more complicated than these simple divergent views might suggest. First, crude measures of total carbon or total nitrogen in sediments do not usually give a reliable guide as to what is available to deposit feeders. A variety of studies have suggested that only something like 5–30% of the organic matter present can actually be utilized. Partly, at least, this is because most deposit feeders are selective: they may pick up a specific size fraction, or some other portion of the sediment. The bivalve *Nucula proxima*, for instance, selectively ingests the organic fraction, and the faeces actually contain a higher proportion of POM than the sediment. This species also actively selects bacteria (Lopez *et al.*, 1989). Selectivity will be discussed later (p. 82).

Second, the detrital picture in intertidal sediments is complicated due to the presence of large populations of photosynthetic microbes, especially diatoms. Microalgae, mostly diatoms, can make up 70% of the organic carbon in some size fractions, while bacteria usually constitute no more than 2%. While sub-surface feeders such as *Clymenella* have only limited access to diatoms, surface feeders may ingest large quantities of them. Experiments on the snail *Hydrobia ulvae* show that when it is maintained on sediment containing primarily bacteria, it increases in weight by nearly 60% in 6 weeks. On sediment rich in diatoms it increases by over 170% in the same time (Jensen and Siegismund, 1980).

Third, there is the question of nitrogen supply for deposit feeders. Most animals need a diet in which the C/N ratio is less than about 17. Fresh detritus often has a value of about 20, and this falls as bacterial populations rise (see p. 19). The bacterial protein may therefore be essential for detritivore growth. On the other hand, C/N ratios in microalgae are often about 6, so these may take the place of bacteria near the sediment surface.

Lastly, the significance of DOM for some species has now been recognized. There is a surprising amount of DOM in the sea—possibly 10 times as much as POM (Fenchel and Jørgensen, 1977). Although much of it may be 'refractory', i.e. in the form of complex molecules not easily metabolized by animals, concentrations of smaller molecules such as amino acids and sugars may reach levels in the μM range (Manahan, 1990). Many soft-bodied marine invertebrates, including larvae, can take up these small molecules through their epidermis, and studies on the polychaete *Clymenella* suggest that a considerable fraction of the worm's energy supply could be derived in this way. Deep-sea phyla such as Pogonophora, which have no gut, as well as gutless species from some other phyla such as the Nematoda, also depend to some extent upon uptake of DOM; but they combine this with energy derived from sulphide metabolism via bacterial symbionts, in a fashion similar to the bivalve *Solemya* (p. 72).

Overall, deposit feeders have to cope with organic food that is generally low in quality and mixed with 95% or more inorganic material, unlike suspension feeders which may take in high-quality food with much less inorganic bulk. Calculation of energy budgets suggests that bacterial biomass is usually too low to support deposit feeders totally, and they depend more upon non-living POM (Heip *et al.*, 1995). However, intake of POM is topped up by feeding on bacteria, diatoms, and DOM, and the proportions of these taken may vary widely.

Who eats detritus and how?—the detritivores

Benthic detritivores are remarkable for their versatility of feeding methods. Very few seem to stick to one approach, several have more than two ways of obtaining food, and as we shall see, the distinction between deposit feeders and suspension feeders is often very blurred.

Pseudopolydora kempi is a 'sedentary' polychaete found on the Pacific coast of the USA, and common where the sediments show ripple marks. Observations in a laboratory flume show that *Pseudopolydora* acts as a deposit feeder on the sheltered lee side of ripples, pressing its palps close to the sediment. As the ripples migrate across the bed, *Pseudopolydora* is left on the exposed side of the ripple face, and then it curls its tentacles into helical coils and becomes a suspension feeder (Lopez *et al.*, 1989). This story is mimicked by the European estuarine bivalve *Scrobicularia plana*. At low tide, when the mud surface is wet, *Scrobicularia* protrudes its inhalant siphon from the sediment and uses it to vacuum up loose detritus from the sediment surface. The siphon creates a characteristic star-shaped pattern because it is extended along different radii from the burrow. At high tide, *Scrobicularia* changes behaviour and the inhalant siphon is withdrawn near to the mud surface. The bivalve then draws in water through it and acts as a suspension feeder (Hughes, 1969). *Macoma balthica* shows the same alternation of feeding strategies, while the west-coast American bivalve *Macoma nasuta* shows even more variety. It changes from deposit feeding in slow water currents to suspension feeding as currents increase. At very fast flows, its siphon may act as a trap for material moving as bedload (Lopez *et al.*, 1989).

The estuarine ragworm *Nereis (Hediste) diversicolor* has so many modes of feeding that it is most properly termed an omnivore. Judging from its gut contents, it is primarily a detritivore, emerging from its burrow to take 'gulps' of surface sediment. When the sediment has a diatom coating, the feeding pattern is clearly seen as a series of irregular furrows, so diatoms may figure largely in the diet. Meiofauna such as copepods and nematodes are frequently eaten along with sediment. *Nereis* can also be a carnivore or scavenger, taking the amphipod *Corophium* and the bivalve *Macoma*, and a herbivore, eating the macroalga *Ulva*. Besides this varied diet, *Nereis* can also suspension feed. To do this, it secretes a mucus net inside the burrow and pulls water through it by undulating the body. Small particles of detritus are caught in the net, which is then eaten, and a new net is secreted (Riisgård and Larsen, 1995).

Mudsnails of the genus *Hydrobia* commonly reach densities of many tens of thousands per square metre in a variety of sediment types. How do they adjust their feeding mechanisms to suit both mud and sand? When *Hydrobia ulvae* is allowed to feed on sediment which has been ^{14}C-labelled, the feeding rates on various size fractions can be measured. Snails feed most rapidly on fine sediments (less than 80 μm), but still take up the radioactive label from the coarser fraction, even when particles are as large as 1000 μm. Fine sediments, including diatoms, are swallowed whole, while coarser particles are taken into the mouth, scraped by the radula, and then spat out—a process christened 'epipsammic browsing' (Lopez and Kofoed, 1980). The snails can also graze on solid surfaces, and can feed on macroalgae such as *Ulva*. Occasionally, they float upside-down on the water surface, and can then feed on the microbial film that collects there.

Crustaceans usually employ appendages with spines or hairs to sort sediment while deposit feeding. The amphipod *Corophium* provides a good example of diverse feeding. It can sift detritus with its gnathopods (mouthparts) while moving over the mud surface. It can filter fine particles from its respiratory current while in its burrow. And it can use its very long second antennae as rakes to seize detrital clumps and drag them back into the burrow. These lumps are then winnowed by the respiratory current, and fine particles are sieved out by hairs on the gnathopods (Meadows and Reid, 1966).

These examples show that many 'deposit feeders' have very varied feeding mechanisms, many of which cannot be distinguished from suspension feeding. While some suspension feeders show similar flexibility, especially on exposed shores (p. 51), the majority do not show much variation in their mechanisms of food gathering. Instead, they specialize in selective mechanisms which operate on particles after they are caught. Many feed primarily on phytoplankton, and mechanisms to capture plankton do not differ greatly from those found on sandy shores. But many muddy-shore bivalves feed primarily on suspended POM. They have selection mechanisms in the gills, where particles are first trapped; in the labial palps, which are organs sorting out which particles are taken into the mouth; and in the gut itself, where selective digestion sorts out what is absorbed. The deep-burrower *Mya*, for instance, can reject inorganic particles at the gills, and can therefore improve the quality of the material ingested. As the organic content of the food decreases, however, the ability to reject decreases, and the clams take in a wider spectrum of particles (Bacon *et al.*, 1998).

Competition

With so many species eating detritus, is there enough food to go round? Or do animals compete for a limited supply? This question is still the subject of intense debate, made particularly difficult by the fact that different species utilize various fractions of the detritus—bacteria, diatoms, and dead POM. Some consideration of microbial kinetics allows calculation of the rate of 'resource renewal'; i.e. the rate at which microbes can replace their populations (Levinton in Livingston, 1979), and therefore should allow an estimate of whether species have to compete for the resource. But some species may not rely upon microbes to any extent. And in any case, it is possible that predation could keep the populations down to a level at which they do not need to compete (Reise, 1985).

Another problem is that even if competition can be demonstrated, it is hard to identify the resource for which the animals are competing. Are their interactions to do with food, or with living space? The sediment is very much a three-dimensional environment with room to escape from neighbours, but as we have seen in Chapter 2, some animals can alter the sediment environment on a massive scale, causing indirect effects on other species. Despite these problems, competitive effects in mudflats have been demonstrated (Wilson, 1991), and are probably more widespread in these deposit feeding communities than in the suspension feeding communities of sandy shores.

On muddy shores in the Baltic Sea, two of the most abundant macrofauna are the annelid *Nereis diversicolor* and the amphipod *Corophium volutator.* In the field, some patches of mud are dominated by *Nereis*, others by *Corophium.* Is this complementary distribution due to competition? When the two species are placed together in aquaria, *Nereis* actively burrows, destroying *Corophium* tubes and decreasing *Corophium* feeding time. As *Nereis* density increases, *Corophium* mortality increases too (Fig. 4.11). More *Corophium* leave their tubes, and if not allowed to establish new ones they become exhausted and may occasionally be eaten by *Nereis*. The competitive 'disturbance' effect is, however, thought to be the major force governing distribution of the two species (Olafsson and Persson, 1986).

Mudflats on San Juan Island, Washington (west-coast USA), have large numbers of polychaetes. Some, like *Armandia brevis*, burrow freely in the sediment while others, like *Platynereis bicanaliculata*, build tubes and remain in one place. The two species compete for space, as shown by exclusion experiments. Mesh cages over the sediment quickly grow a diatom film, and these were used to exclude the settling larvae of the tube-builders, leaving the larvae of the burrowers to penetrate the cages and then to burrow in the sediment beneath. With cages in place, density of *Platynereis* tubes declined to less than 20% of their uncaged density, while burrowing *Armandia* trebled their numbers (Woodin, 1974). Separate experiments showed that however many worms were introduced into a fixed volume of sediment, the numbers rapidly stabilize at a fixed worm volume: sediment volume. No direct actions between the species have

Fig. 4.11 The mortality of the amphipod *Corophium volutator* when placed in aquaria with the polychaete *Nereis diversicolor.* Thirty adult *Corophium* were used, and mixtures of large and small *Nereis.* Bars show SD. (From data of Olafsson and Persson, 1986.) The insert is after a vignette entitled 'Combat between an annelid and *Corophium longicorne*' by C.S.Bate and J.O.Westwood (1863): *A history of the British sessile-eyed Crustacea* (Van Voorst, London).

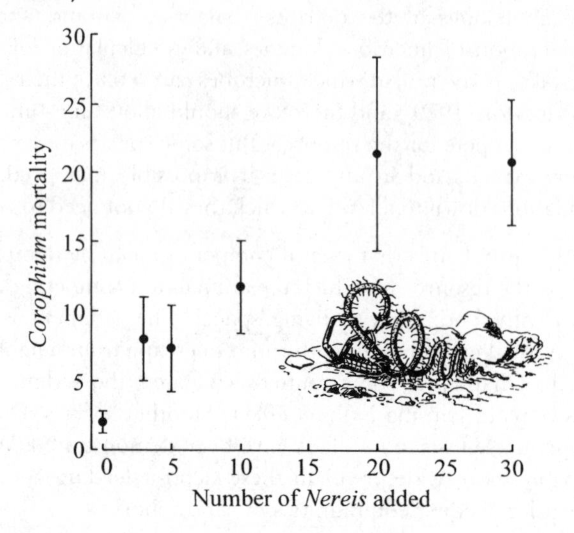

been observed, and the effects are probably indirect: the tube-builders retard the movement of the burrowers and leave less space for them to move around and feed.

Competition *can* thus occur on the mudflats. But does it normally occur? Experiments are seldom able to show competition actually occurring under natural conditions. But there is good evidence that food can be in limited supply. Field experiments in England show that when the snail *Hydrobia ulvae* is confined in small areas at natural densities, mortality and growth rate decline as density increases. On the other hand, these increases in density of *Hydrobia* appear to have no effect upon the amphipod *Corophium* (Morrisey, 1987). Could it be that instead of competing, animals have subtle ways of partitioning food and space, so they avoid competition?

One classic example again concerns the mudsnails *Hydrobia* spp. In a lagoon in northern Denmark, the Limfjord, *Hydrobia ulvae* and *Hydrobia ventrosa* can both be found. In some places they exist together ('sympatric' populations), while in others the populations live separately ('allopatric'). Allopatric populations of the two species show nearly identical shell sizes, from 3 to 3.5 mm. But in sympatric populations, *H. ulvae* are mostly larger than 3.5 mm while *H. ventrosa* are mostly smaller than 3 mm. Why the difference in size? The change in sympatric populations is known as 'character displacement'. Different sizes of snails have been shown to feed on different-sized particles, so the small snails eat different food from the larger ones—and so do not compete with them. Thus character displacement leads to reduced competition (Fenchel, 1975). It may therefore allow species to coexist rather than tending towards extinction of one species by another. Mechanisms like this may help to explain why so many species can coexist on the flats.

However, this may not be the whole story. Experimental studies on *H. ulvae* and *H. ventrosa* have been carried out in a lagoon in England. Here, when *H. ulvae* is introduced into a cage in large numbers, its density declines, presumably due to *intra*specific competition. At the same time, numbers of *H. ventrosa* actually *increase*, giving no hint of interspecific competition. Similar results apply to *H. ventrosa* in high densities: it declines, while numbers of *H. ulvae* increase. Cherrill and James (1987) concluded that while intraspecific competition may be important in the field, there is little evidence that interspecific competition is important outside experimental situations. The subject will evidently provide scope for further investigation.

Predation

There are many types of predator on mudflats. Wading birds, as we have seen, arrive in enormous numbers in the northern winter. Fish and crabs migrate in and out on the tide. The infauna contains numbers of predatory polychaete and nemertine worms. Besides all these, parasites probably abound, although there have been few studies of their importance. In the European mudsnail *Hydrobia*, trematodes can cause gigantism (Gorbushin, 1997), and in this and several other snails they can cause 'parasitic castration'—degeneration of the host's gonads. In the American snail *Ilyanassa*, trematodes may cause the snails to change their behaviour, making excursions upshore and presumably affecting their exposure to predators. The parasites are dispersed by their highly mobile definitive hosts, which are shorebirds and fish (Curtis and Hubbard, 1993). Predators are thus important in more ways than just the direct effects of predation. In their predatory role, two main questions arise. How do they obtain their prey? And what overall effect do they have on the infauna?

Waders usually use one of two major strategies to capture individual prey animals with their fine bills (Fig. 4.12). One group, the sandpipers and oystercatchers, are mainly tactile foragers. They have long bills which have many pressure receptors to help them detect prey when probing in the mud. They may scan a narrow strip of mud visually, then when signs of activity are given by a prey organism—water emerging from a burrow, or tentacles appearing—they peck at the place or probe into the mud, picking up single *Corophium* or *Nereis*. The second group, the plovers, are mainly visual foragers. They have short bills with pressure receptors mainly at the tip. Plovers stand in one spot and scan an area in front of them, then if signs of prey are seen, they run to peck at it. If no prey are evident, they run to a new scanning position and repeat the procedure (Ferns, 1992). Extraordinary numbers of prey can be picked up rapidly by these methods. Plovers frequently catch five worms per minute, and sometimes double this rate. Sandpipers may take more than 10 000 *Corophium* in 1 day!

Fig. 4.12 Two of the feeding strategies used by shorebirds on mudflats. The 'sandpiper' strategy involves visual scanning while walking along, and the birds peck at surface prey and probe for deeper animals. In the 'plover' strategy, the birds stand at one spot and scan the area in front of them before either probing or running to a new scanning position. (After Ferns, 1992.)

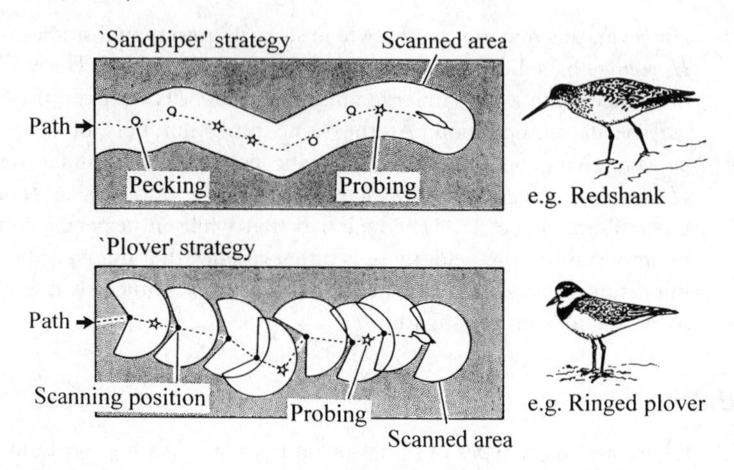

On rocky shores, the effects of predators on the community have been shown by classic exclusion experiments. When dominant predators such as starfish or dogwhelks are excluded, prey species—mussels or barnacles—increase enormously, and outcompete other species so that overall diversity decreases (Little and Kitching, 1996). These experiments show that rocky shore predators can have both powerful direct effects on their prey and indirect effects on the whole community. On muddy shores, the results of predator exclusion are much less dramatic. This is because, as we have emphasized in Chapter 1, predation within sediment is very different in its nature from predation on rock. On a solid two-dimensional surface, prey are often sessile, and even if mobile they cannot easily escape from a predator. In soft sediment, prey organisms can often evade their predators, so predation is unlikely to wipe out a prey species. Indeed, partial predation (e.g. cropping of bivalve siphons by flatfish) is common, and in this case the prey may survive to feed the predator again. In general, the removal of predators from a muddy shore results in a gradual increase in abundance of infauna, but no rise of one particular species.

The most obvious predators on muddy shores, and the easiest to exclude, are birds, crabs, and fish—the so-called epibenthic species. On mudflats in southern California, exclusion of wading birds and fish has significant effects on the infauna in the winter, allowing benthic polychaete numbers to increase. However, species diversity of the infauna is not affected, and the effects of predators are relatively mild. In particular, benthic-feeding fish have less effect than the waders (Quammen, 1984).

Other studies have emphasized the point that because of the complex interactions between species on muddy shores, removal of predators often produces complex effects. On the intertidal flats of the Wadden Sea, north Germany, cages with a mesh size of 2 cm which keep out flatfish, wading birds and gulls, have different effects on different sizes of infauna. Small species such as *Hydrobia* become less abundant. Larger species such as the crab *Carcinus*, large *Macoma*, and the snail *Littorina littorea* increase in numbers. Reise (1985) argued that these changes can be explained by changes in predation: the epibenthic predators prey mostly on the larger infauna, so these increase when predators are excluded. As some of those that increase in abundance are themselves predators, however (e.g. species such as *Carcinus*), they consume the smaller macrofauna, which consequently decline.

On an intertidal flat in Maine, eastern USA, the effects of epibenthic and infaunal predators can again be seen to be interlinked. Experimental exclusion of birds, fish, and crabs results in a slow increase in total infauna: after 20 weeks, densities were 1.5 times higher in cages than in untreated areas. Why this slight effect? As in the Wadden Sea, large predators probably operate via their effect on infaunal predators such as the polychaete *Nereis virens*. When they are excluded, *Nereis* multiplies, preying on many infaunal species and preventing numbers from rising. But *Nereis* also has an infaunal predator, another polychaete *Glycera dibranchiata*. If *Glycera* numbers are high, it both preys on *Nereis* and disturbs it, so *Nereis* numbers fall, and other species increase (Ambrose, 1984).

Similar experiments undertaken over a period of 5 months in tidal creeks of a salt marsh on east-coast USA showed that both infaunal density and biomass rose when predatory fish and crabs were excluded. Increase in biomass was primarily due to a rise in numbers of large predatory polychaetes such as *Nereis* and *Glycera*. The effects of these predators, like those hypothesized by Reise in the Wadden Sea, is to prevent other infauna from increasing greatly (Sarda *et al.*, 1998).

There is at least some uniformity in the reactions of soft-sediment communities to the removal of predation pressure (Peterson in Livingston, 1979). But the story is different if we turn to the communities in seagrass beds. Exclusion cages placed here seem to have little effect on infaunal abundance, suggesting that the seagrasses already offer a refuge for the infauna (Wilson, 1991). And experiments on some estuarine shores have found little effect of predators. In the Ythan estuary, Scotland, exclusion cages have few consequences for the infauna, possibly because the dominant infaunal species there are more mobile than those on flats studied by others (Raffaelli and Milne, 1987).

These few examples serve to show that besides a great variation in the direct effects caused by epibenthic predation, there may often be many indirect effects. In many cases these may explain the otherwise paradoxical effects of exclusion experiments (Kneib, 1991). Thus the effects of predators on muddy shores are usually far more subtle than those on rocky substrata. Even on mudflats, though, dramatic cascade effects can sometimes occur. In the Bay of Fundy, the arrival of more than 100 000 migratory shorebirds caused a decline in abundance of the amphipod *Corophium*; and those surviving showed different behaviour, staying below the sediment surface. The diatoms at the mud surface therefore increased, and as they produced more polysaccharide secretions, the mud surface increased in stability (Daborn *et al.*, 1993). The arrival of predators thus indirectly resulted in increased sediment strength.

Food webs and energy flow

Invertebrate biomass on intertidal flats can be high—in the Chesapeake Bay it can reach over 10 gC/m^2, while in the Ythan estuary (Scotland) it may be up to 30 gC/m^2. Rates of annual production may be twice these values (Heip *et al.*, 1995). In comparison, production rates on the offshore continental shelf are only 2–3 gC/m^2 per year. On the flats, much of the biomass consists of detritivores. Why are mudflats able to support such high levels of detritus-consuming animals? Figure 4.13 shows some of the sources and sinks of carbon on a mudflat. There are no cases in which all these have been quantified, but we have figures for some of the sources. It is, for example, possible to measure production by benthic diatoms (30–230 gC/m^2 per year) and by seagrasses (150–600 gC/m^2 per year). But it is much more difficult to assess how much POM arrives on the flat from elsewhere, or how much carbon is taken by predators. The lower levels of the food webs are particularly badly understood—as we have just discussed, it is not even certain how much detritivores rely upon dead POM and how much they eat living microbes. Recent measurements of consumption rates by protists suggests that these also might be very important in food webs: in fine sediments of the Clyde estuary, Scotland,

Fig. 4.13 An outline carbon budget for an estuarine mudflat. The sources and sinks have not all been quantified on any one flat. DOM, dissolved organic matter; POM, particulate organic matter. (Modified after Heip *et al.*, 1995.)

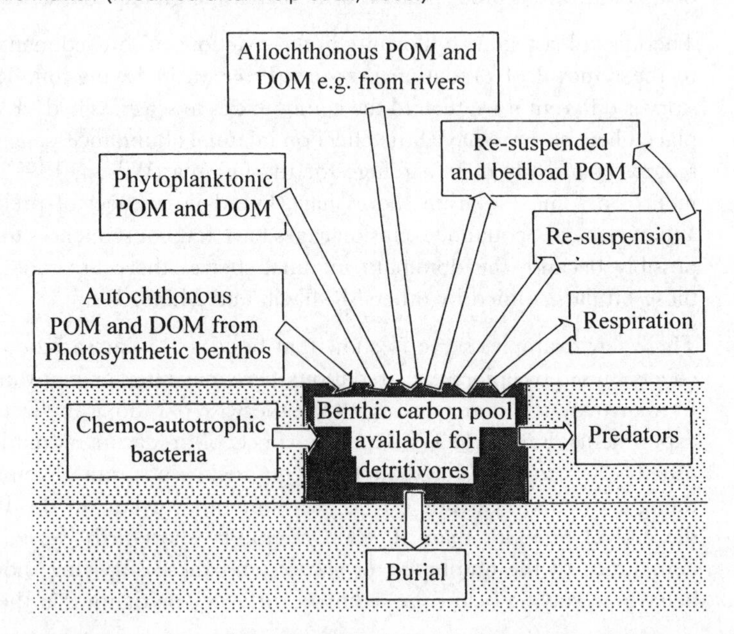

amoebae may remove only about 0.5% of the standing stock of bacteria per hour, but in some circumstances they can utilize 6.5%/hour (Butler and Rogerson, 1997). There are therefore many questions still unresolved, and we consider only two—the source of detritus for detritivores, and the importance of predators in exporting carbon from the flats.

Detritus can be derived from seagrasses, benthic diatoms, phytoplankton, salt marsh grasses, or from POM that has more distant origins. Is it possible to tell which kind of detritus an organism is eating? The use of stable carbon isotopes as 'tracers' has allowed estimates to be made for some benthic invertebrates (Mann, 1988). The ratio of ^{13}C to ^{12}C in plants is usually expressed as the $\delta^{13}C$ ratio, and this varies with different plant types. For instance, phytoplankton has values about −20 to −22‰, while in benthic diatoms the values are −16 to −18‰, and salt marsh grasses are −12 to −14‰. Measurement of the $\delta^{13}C$ ratio in animals gives a clue to the origin to their food source. In salt marsh creeks, for example, most invertebrates have $\delta^{13}C$ ratios more negative than −15‰, indicating some input from benthic diatoms and phytoplankton as well as from salt marsh grasses. The use of other isotope ratios such as ^{15}N and ^{34}S can help to pin down exact sources for detritus, and this 'tool' should be of great use in refining food webs in the future.

We come now to the importance of predators in exporting material from the flats. There are many problems here. Estimates may be made by comparing production or biomass within exclusion cages with values for surrounding mudflat, but it is difficult to be sure that cages do not create artificial conditions. On the other hand, direct calculation of how much food birds or fish take is not easy. Direct observation may give a good idea of the intake rate for birds, but this is difficult at night, and in any case it is hard to estimate the size or weight of individual food items through a telescope! Taking these factors into account, estimates of the effects of predators are necessarily approximate (Baird *et al.*, 1985).

In the Wadden Sea, Europe, birds have been calculated to consume nearly 17% of annual invertebrate production, taking both intertidal and subtidal areas into account. Fish may take another 17%, while crabs take 10% and fisheries for cockles and mussels a further 7%. Altogether, 57% of production may be removed by predators. Is this a typical value? In the Ythan estuary, Scotland, birds may take as much as 36% of annual production, while in the Langebaan Lagoon in South Africa, the figure may be only 20%. Despite this variation, it is evident that a very substantial proportion of production may be removed, and predators may have significant effects upon the flats.

Techniques

Access to the mudflats, and sampling

One of the major problems in trying to investigate mudflat organisms is access to the flats. While some mudflats can be visited on foot, many are deep and treacherous. Individuals should therefore always work in groups of two or preferably more. Note, however, that tidal flat communities are vulnerable to damage from too much digging or trampling (p. 212), so in the interests of conservation investigations should not involve intensive sampling by large groups. When normal access on foot is possible, the sampling methods outlined for sandy shores (p. 56) can be used. When this is impossible, various techniques can be employed. Equipment may be carried on a sledge, thus lightening the load on the observers. If access is required only over short distances, temporary walkways can be constructed from materials such as wooden fencing. Walkers should, however, beware that on some long flats, the tide can move across the mud faster than a person can walk. Strict attention to the tide tables can obviate disaster.

On many flats, the only practical solution is to take samples from a boat at high tide. Various forms of grab can then be used, depending upon the sample size required (Holme and McIntyre, 1984), but deployment from a boat may mean that plotting the positions at which the samples were taken is difficult.

The extraction of fauna from mud samples is also more difficult than from sand. A sieve of 0.5 mm mesh is normal for macrofauna, but mud often forms resistant balls when attempts are made to sieve through this fine a mesh. Samples are best preserved in formalin before sieving, and very gentle use of the sieve, combined with gentle water jets, helps to separate the animals from the sediment.

The use of exclusion cages

Cages to exclude predators can only be successfully employed in areas where tidal currents are not too fierce. Their design has a series of inherent problems (Olafsson *et al.*, 1994). Even with large mesh sizes, they may alter water velocities and therefore change the sedimentary regime, leading to deposition of small particles or to scour. It is essential to include variants like partial or open cages in order to allow for cage effects. It is equally essential to realize that these may not really be adequate controls (Hulberg and Oliver, 1980).

Observation of mobile predators

The fact that fish and crabs invade the flats only as the tide rises means that observing them is virtually impossible at most sites. Nevertheless, seine nets can be used to sample fish populations at different stages of the tide, and a variety of other techniques for catching fish can successfully be employed (Baker and Wolff, 1987). Calculations of fish abundance on the flats are usually difficult.

Observation of birds is a good deal more practicable. One of the most informative routine methods has been to make monthly counts of birds, usually at their roosting sites. This method provides an assessment of how 'important' any one site is for particular species. Coupled with ringing schemes, these counts allow much insight into the use that waders, gulls, geese, and others make of various sites (Ferns, 1992).

5 Salt marshes and mangrove swamps

On their upper tidal levels, most marine and estuarine mudflats are hidden by vegetation. In temperate zones, salt-tolerant grasses and other herbaceous plants form what is called salt marsh. Here in spring and summer the white flowers of scurvy grass or the yellow of asters and the purple of sea lavender form colourful swathes against a background of cordgrass, while in winter the vegetation dies back to brown remnants. In the tropics, salt-tolerant woody plants called mangroves form considerable forests, mostly evergreen, known either as mangrove swamps or mangals. Both salt marshes and mangrove swamps are areas of very high productivity and are home to enormous diversities of animals.

Distribution of salt marshes and mangrove swamps

Plants do not occur on every soft shore, and their distribution is determined by factors such as shelter, slope, and sediment supply. Salt marshes and mangroves occur mostly where shores are sheltered from wave action, for instance behind shingle bars, in lagoons and estuaries. But occasionally, where wave energy is relatively low, as in the Florida 'panhandle', mangroves grow on unprotected shores, while salt marshes occur on unprotected shores on the east coast of Britain. Development is best where the shores are gently sloping, and where there is a good but not excessive supply of sediment. Salt marshes are most extensive where there is significant tidal amplitude, but mangroves are often found where tides are very small.

The division between salt marshes and mangroves is clearly related to temperature. Mangroves are most common and diverse where mean temperature in the coldest month does not fall below 20°C, and they do not occur where the mean monthly temperature is below 10°C. Nevertheless, this allows them to grow as far south as North Island, New Zealand, and as far north as Japan. Salt marshes take over where mangroves are absent, and spread well into the Arctic circle (Fig. 5.1). They are not absent from the tropics, however, and can often be found in patches mixed with mangroves.

Why is it, then, that mangroves are temperature-limited? Present-day mangrove species belong to a wide range of unrelated families, and all of these families are primarily found in the tropics and sub-tropics. Yet there is nothing obvious about their physiology that limits them to high temperatures (Adam, 1990).

Fig. 5.1 The distribution of salt marshes and mangrove swamps. Species numbers show maximum mangrove species in the area. (After Trenhaile, 1997, and Stafford-Deitsch, 1996.)

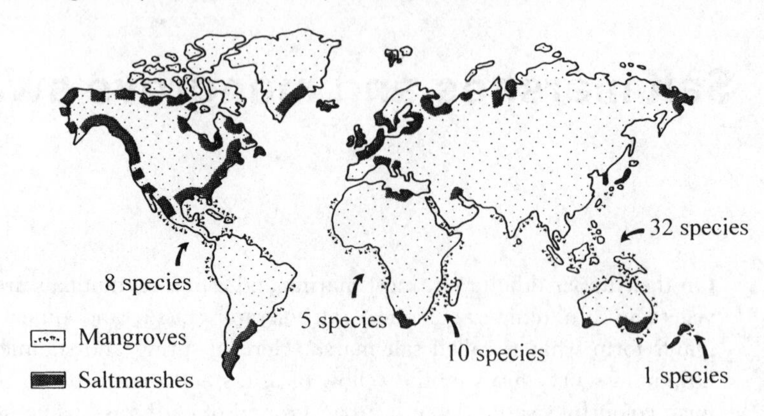

One suggestion is that it might be easier to cope with salt stress at high temperatures. A salty environment certainly puts the plants under stress, as many species grow better in dilute than in full-strength sea water, and experimentally many species can be grown in fresh water. Alternatively, there may be an indirect link here: coping with salt stress may need functional leaves throughout the year, and the evergreen state is easier to maintain and is more widespread in the tropics. Most mangroves are evergreen, and some—but not all—do use the leaves to excrete excess salt.

Salinity is, in fact, another major determinant of the distribution of salt marshes and mangroves. Where salinity is above 5 (5‰ in older terminology—see Chapter 8), the plants are mostly salt-tolerant and are called halophytes. Where salinity is less than 5, halophytes are mostly replaced by freshwater plants. Thus salt marshes and mangrove swamps are replaced by other species at the freshwater ends of estuaries. As we shall discuss later, however, this replacement is due more to competition with freshwater plants than to any physiological inability to withstand fresh water. At the very top of the shore, where tides no longer penetrate, halophytic vegetation is normally replaced by terrestrial vegetation, but in many arid regions mangroves may give way to desert halophytes.

In the last chapter, we discussed the seagrasses, which live low on the shore and require prolonged water cover. Salt marshes and mangroves, in contrast, resemble the majority of other angiosperms in that they cannot for the most part withstand long periods of submersion. As a rough guide, the lowest species are found at about the level of mean high water neaps (MHWN), and are not covered by about 10% of the approximately 700 high tides in a year. Many species are found much higher up and may seldom be covered by the tide at all. But this 'rule' is far from universal. In extreme shelter, salt marsh grasses in Britain may be found down to mean low water neaps (MLWN). In North America, the salt marsh grass *Spartina alterniflora* seems more tolerant of submersion, and is regularly found down to mean tidal levels (MTL). In the Indo-Pacific and in Australasia, mangrove genera such as *Avicennia* are normally found only above MTL. But in the Caribbean, the genus *Rhizophora* may be found down to low tide level, where the roots are continously submerged. Overall, then, coverage by water is only one factor that interacts with other physical determinants.

Formation and morphology

Mature salt marshes and mangrove swamps are dominated by a branching pattern of drainage creeks, with higher vegetated areas between, and sometimes with isolated lagoons or pans. The creeks serve to drain water off the vegetated flats as the tide falls, but are also important in channelling the flood water as the tide rises. How do such systems arise in the first place?

We start with a temperate-zone salt marsh as an example. A marsh will form only where sediment is already accreting, although marshes may go through cycles of erosion as well as accretion. Up to the level of MHWN, mudflats may be stabilized by microalgae such as diatoms (p. 25). Accretion may also be encouraged by the filamentous alga *Vaucheria* which forms a moss-like coverage. But when a mudflat reaches MHWN, angiosperms may begin to colonize. Genera such as *Salicornia*, the glassworts (Fig. 5.2), and *Spartina*, the cordgrasses (Fig. 5.3), have roots that bind sediment particles together, while their stems reduce water velocity and encourage further sedimentation. Growth of vegetation is usually patchy, so the salt marsh develops as a series of uneven hummocks, with depressions and drainage areas between.

Fig. 5.2 Section showing the general physical features of a salt marsh. Diagrams at the top show some structures in plan view. At the bottom, plants typical of a marsh in southern England are shown in their characteristic regions.

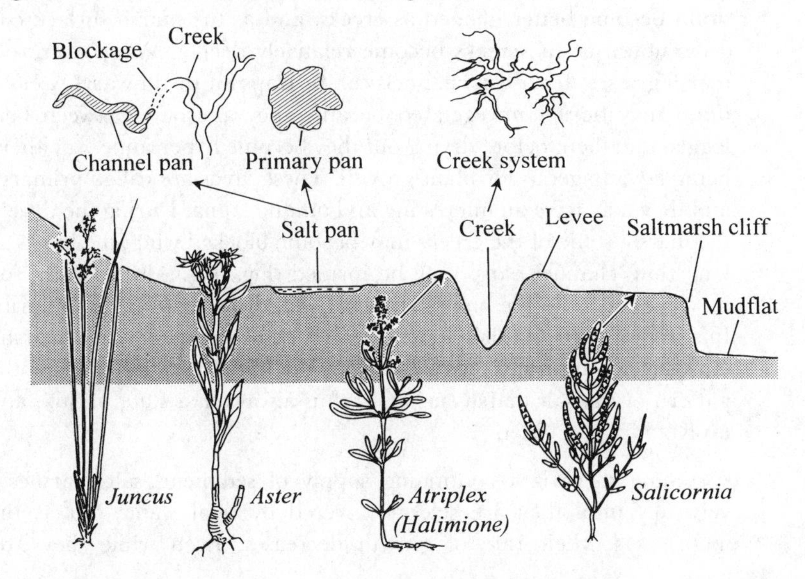

Fig. 5.3 Section through a salt marsh in the Severn estuary, UK. The marsh was laid down in three successive phases, the first two leaving 'fossil' salt marsh cliffs. Distribution of four grasses, *Elymus (Agropyron) pycnanthus, Festuca rubra, Puccinellia maritima,* and *Spartina anglica* is shown below. MHWN, mean high water neaps; MHWS, mean high water springs. (After Little, 1992, and Allen and Pye, 1992.)

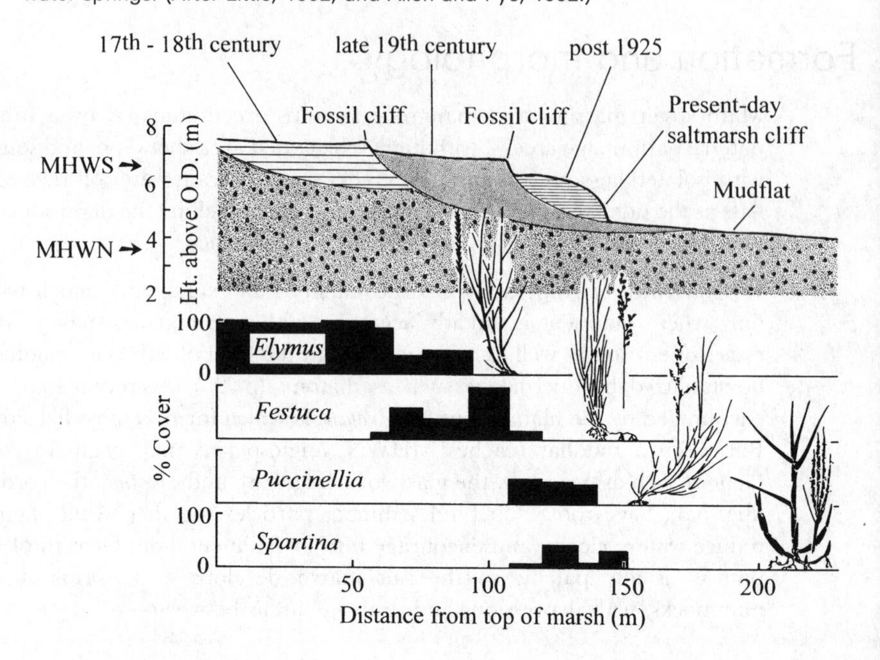

As the marsh grows in height, conditions become suitable for other species to grow, including a number of broadleaved herbaceous plants such as *Limonium*, the sea lavenders and *Aster*, the sea asters (Fig. 5.2). Meanwhile the areas that drain become better defined as creeks, and as the marsh surface continues to grow upwards, the creeks become relatively deeper. When overgrown by salt marsh grasses, 1 m deep channels can be traps for the unwary! Areas that do *not* drain may be left unvegetated because they alternate between being waterlogged and then, before drying out they become hypersaline—neither condition being advantageous for plant growth. These areas are called primary salt pans, and they may have an interesting and unique fauna. During the development of the marsh, some of the creeks may become blocked when the banks slump, and long thin 'channel pans' may be formed (Fig. 5.2). Alternatively, some creeks may cut back into the marsh at their head, draining some of the primary pans. In addition, rafts of tidally-borne litter may be dumped on the marsh and these may develop into unvegetated pans as the initial vegetation is smothered. The pattern of the salt marsh surface is thus always in a state of flux, and tends to evolve as it gets older.

Assuming there is a continuing supply of sediment, salt marshes will grow vertically until they are seldom covered by tidal water, but as their height increases so their rate of growth decreases. Even while they are growing vertically, however, they may suffer erosion at their seaward edge, particularly if wind or tidal conditions change. Quite commonly, a small erosion cliff may be formed—the 'salt marsh cliff'. We then have a situation where the lower part of the marsh may still be accreting vertically, while being cut back horizontally. As the cliff retreats, some of the eroded material is dumped at its foot, and secondary marsh may then form here. Such changes in the erosion climate may recur repeatedly, resulting in a series of cliffs at various heights on the marsh. In the Severn estuary, for example, three cliffs are found, each at the front of a marsh flat, so the marshes consist of three terraces of different ages (Fig. 5.3). The oldest and highest was formed in the late seventeenth century, the middle one in the late nineteenth century. The lowest terrace commenced after about 1925, but is already retreating in some places at 1 m/year (Allen and Pye, 1992).

Mangrove swamps form in a similar fashion to salt marshes, in areas of accretion. Roots and stems aid the trapping of sediment, and where the sediment supply is large, the swamp can advance at enormous rates. In Java, some have been shown to advance at 200 m/year. Mangroves contribute leaf litter as well as trapping allochthonous sediment (i.e. sediment originating elsewhere), and can also raise the height of the swamp rapidly.

Distribution of flora and fauna

The plants that make up marshes and swamps, and the animals that live within them, vary widely both with latitude and from continent to continent. Here is a brief account of the distribution of some of the major genera.

Salt marshes

In the Arctic, salt marshes contain very few plant species, probably because they are subject to ice formation each winter which may prevent long-term successions. Dominant species in the Arctic are the grasses, particularly *Puccinellia*. Coming south into boreal zones, various halophytic herbs such as *Triglochin* (arrow-grass) and *Salicornia* (glasswort) are added to the flora. In temperate zones, diversity increases, and individual regions are characterized by particular genera. Thus South Africa and Australasia have glassworts called *Sarcocornia* instead of *Salicornia* (Underwood and Chapman, 1995). On the Pacific coast of North America, the dominant grass is *Distichlis*. In Japan, China, and Hong Kong, a grass called *Zoysia* forms high-level salt meadows. It is often found above the mangroves, colonizing sandy rather than muddy substrata, and it produces distinctive communities (Morton and Morton, 1983). Cordgrasses (*Spartina*) are widespread: on the Atlantic coast of the USA, *S. alterniflora* takes over the low marsh while *S. patens* covers high marsh. In Europe, *S. anglica* is abundant, but in places the marshes have what is sometimes called a 'general salt marsh community'—an association of broadleaved herbs. In the tropics, *Spartina* spp. also occur, but botanical diversity in the marshes is reduced again, as in the Arctic.

Fauna on salt marshes consists of three groups. First there are the specialists, and we shall concentrate on these. But the specialists are mixed with those species that use the marshes as extensions of the mudflat habitat, and occupy particularly the creeks, the pans, and the bare areas of mud between plants. In addition, many terrestrial species venture downshore and use the marsh as an extension of the land.

Salt marshes are areas of very high plant biomass, so as one might expect there are enormous numbers of phytophagous insects. Sap-suckers such as aphids and bugs, and nectar-feeders such as butterflies and dipterans are abundant among and above the stems of the grasses and herbs (Fig. 5.11). In American marshes, some of the dipterans are the first to make your acquaintance: while the males of horse flies (*Tabanus*), deer flies (*Chrysops*), and mosquitoes (*Aëdes*) feed on nectar and pollen, the females are vicious biters. Also in the plant canopy of most marshes is a great diversity of spiders. These vary from mobile hunters to sedentary web-spinners, the Linyphiidae (Heydemann in Jefferies and Davy, 1979).

On the salt marsh surface, beneath the vegetation, the invertebrate fauna consists of detritivores: more insects—mainly beetles—and a mixture of snails and crustaceans (Fig. 5.11). Many of the snails are pulmonates (air-breathers): in Europe and west-coast USA *Ovatella*, in east-coast USA *Melampus*, and in Australia *Salinator* among others. East-coast USA has many crabs, from fiddler crabs like *Uca* to grapsids like *Sesarma*, but this is unusual. Most marshes have only small crustaceans like the amphipod *Orchestia* and the isopod *Sphaeroma*. East-coast American marshes are also unusual in having large numbers of suspension feeding mussels, *Geukensia*.

The upper levels of salt marshes may be visited by a wide variety of vertebrates. Mammals such as foxes and otters prey upon the invertebrates, while a range of grazers from kangaroos to voles takes advantage of the herbage. Birds of prey, ducks, and waders also visit, but waders use the marshes more for roosting sites than for feeding or nesting. Even then, they seldom use marshes that have been left untouched by humans, preferring areas grazed by cattle or sheep where visibility is greater.

Mangrove swamps

There are about 50 species of mangrove trees, and they can be grouped into two main areas of diversity. The major one centres on the Indo-West Pacific, from eastern Africa through Indonesia to Australia. Here *Avicennia*, recognized by its aerial roots or pneumatophores that stick up through the mud like pencils (Fig. 5.4), is typical of the seaward edge of the swamp. Sometimes it is mixed with or replaced by another genus, *Sonneratia*, and occasionally it occurs at the landward edge as well. Behind *Avicennia* comes a region in which tangled supporting 'prop' roots of *Rhizophora* may make penetration of the swamp almost impossible. *Rhizophora* may be mixed with other genera such as *Ceriops*

Fig. 5.4 Mangrove root systems. *Avicennia* has a radial system of shallow roots that give rise to pencil-like pneumatophores. *Rhizophora* is characterized by arching prop roots. *Bruguiera* has knee-shaped pneumatophores that often give rise to secondary roots.

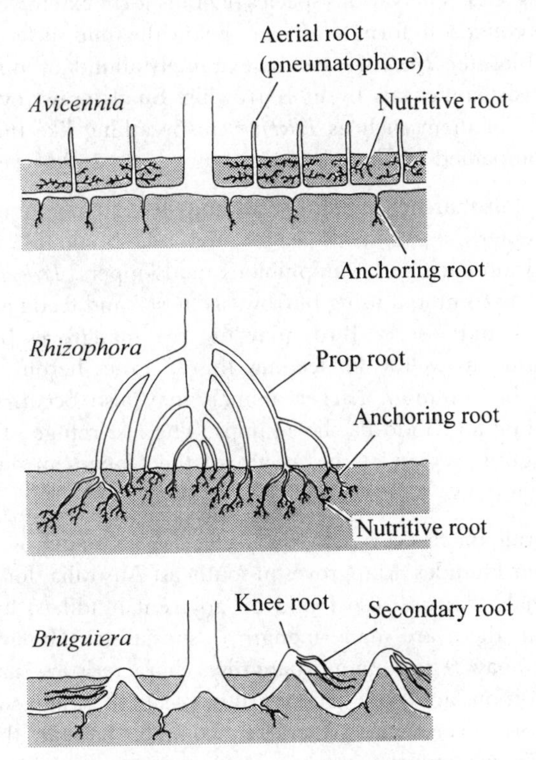

and *Bruguiera*. Curiously, this distribution tends to be reversed in the other area of mangrove diversity—Atlantic coasts and the Pacific coast of America, the most diverse region being the Caribbean. Here *Rhizophora* grows at the seaward edge of the swamp while *Avicennia* is found more in the landward part.

Mangrove swamps offer a wide variety of niches for animals to inhabit—more so than salt marshes because mangroves provide all the microhabitats of a forest. In the tree canopy, an essentially terrestrial fauna flourishes—insects such as ants and termites are abundant, and there is a great diversity of spiders, the most obvious being the orb spiders that construct huge nets between the trees. Specialist leaf-eaters even include the proboscis monkeys that swing through the trees in Borneo. As in salt marshes, many of the insects flying through the swamp are unpleasant to humans. Mosquitoes are rife, and may be accompanied by the biting midges or Ceratopogonidae.

Mangrove trunks and pneumatophores offer a hard surface in an area otherwise dominated by mud (Fig. 5.9), and may be thickly covered by oysters and barnacles. Snails of the genus *Littoraria* crawl over the trunks and leaves—one species in the Atlantic but 20 in the Indo-Pacific. Some of the crabs are also arboreal—*Aratus* in the Caribbean, *Sesarma* and *Metopograpsus* in the Indo-Pacific—but most live on or in the floor of the swamp, together with the fiddler crabs, *Uca*. The various species of crabs form extensive burrows, many of which interconnect to form 'crab runs' below the mud surface. In the Indo-Pacific, the mud-lobster *Thalassina* can be extremely abundant in mangrove swamps, creating vast upheavals by its burrowing. Snails crawl over the mud surface too, many of them such as *Ellobium* air-breathing like those in salt marshes, and accompanied by slug-like pulmonates such as *Onchidium*.

Like salt marshes, mangrove swamps have many vertebrate visitors. Shoals of fish enter the swamps to feed at high tide. Some fish, however, are resident. In the Indo-Pacific, the amphibious mudskippers, *Periophthalmus* and its relatives, may be common, using burrows as 'nests' and feeding on both marine invertebrates and insects. Birds may use the mangroves both for nesting and for roosting, as well as for feeding. Ibises, egrets, herons, pelicans, and kingfishers may be common. Larger animals may also occur. Crocodiles inhabit both Atlantic and Indo-Pacific swamps. The last refuge of the Bengal tiger is in a mangrove region at the mouth of the Ganges in northeast India called the Sundarbans.

Overall, the fauna is more diverse in tropical mangrove swamps than in those of higher latitudes. Mangroves in southeast Australia, for example, contain mostly animals that are also found on adjacent mudflats; but in mangroves further north, there are many mangrove 'specialists' (Hutchings and Recher, 1982). This may reflect many mangrove characteristics that change with latitude. Conditions are more constant in tropical areas, and swamps there also provide a more diverse and structurally complex habitat: the number of mangrove species increases from one in southern extremes of Australia to 29 in northern areas. In addition, mangroves cover a greater area in tropical regions than on their northern and southern fringes, and this may have allowed more species to evolve.

Organisms and their adaptations

The conditions in salt marshes and mangrove swamps are critical for the survival of the plants that form them and the animals that live within them. Below we consider some of the ways in which the organisms are adapted to their environment.

How to cope with too much salt

High concentrations of salts are toxic to the tissues of flowering plants, and they may also interfere with the uptake of essential nutrient ions. Yet full-strength sea water contains approximately a 500 mM solution of NaCl, together with substantial quantities of magnesium and sulphate. How do salt marsh plants and mangrove trees cope with this problem?

Most aquatic halophytes can actually exclude ions from uptake by their roots, keeping salt concentrations in the plants below that of the environment. Thus the concentration in the xylem—the transporting stem tissue—is usually no greater than 100 mM NaCl (one-fifth of sea water). The barrier to salt is created by the membranes of the root cells which have a very low permeability to sodium and chloride, and presumably have active transport systems that pump salt outwards—in the arrow-grass *Triglochin maritima*, root cells show an efflux of sodium chloride when external levels are above 10 mM (Flowers *et al.*, 1977). This mechanism is not, however, completely effective, and some ions do enter, so all halophytes have ways of storing or excreting salts. One way of dealing with the incoming salts is to dilute them by developing cells with a very high water content—the so-called 'succulent' tissues. Many halophytes employ this technique—the glasswort *Salicornia*, for example, has succulent stem cells, while the mangrove *Rhizophora* has succulent leaves. A separate technique is to excrete salt through special glands—the salt glands (Fig. 5.5). Halophytes in a wide range of families have glands on the leaves that can actively secrete salt, so much so that dry leaves of the mangrove *Avicennia* often glisten with a white salt layer. This secretion helps to lower internal salt concentrations, but we should note that, although salt glands are taxonomically widespread, the majority of halophytes do not possess them.

Fig. 5.5 Secretion of salt by the cordgrass *Spartina anglica* (open circles) and the sea lavender *Limonium vulgare* (filled circles). The inserts show salt glands in the two species. Although the *Spartina* glands consist of only two cells, they secrete salt much faster than the multi-celled glands of *Limonium*. (After Rozema *et al.*, 1981, and Lüttge, 1975.)

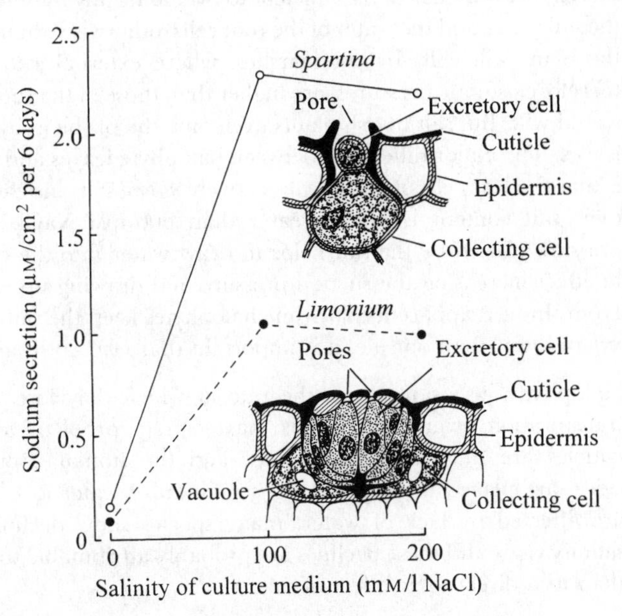

In spite of these mechanisms, halophytes have higher internal salt concentrations than freshwater or terrestrial plants, and physiological adaptation to this situation has been necessary. How do they keep their metabolic systems functioning when the fluids in the cell vacuoles have salt concentrations high enough to inhibit enzyme function? The answer is that the plants keep their cytoplasm at the same osmotic pressure as the cell vacuoles by accumulating amino acids such as proline and betaine, but excluding salts. Thus there is no problem of cells flooding or dehydrating from an imbalance of osmotic pressure, but the enzyme systems can still work in a relatively salt-free environment.

In summary, halophytes keep down their internal salt concentrations to some extent, and then balance the rise that does occur by internal cellular mechanisms. However, by keeping their internal osmotic pressure lower than outside, they incur another problem: effectively, salt marsh and mangrove plants live in a self-imposed physiological desert because water tends to flow out of the plants rather than in. How do they survive in this situation?

In simple terms, plants normally take in water through the roots, then lose it by transpiration through the leaves, and their state of hydration depends upon a balance between these two processes. Perhaps unfortunately, however, no organism can actively transport water by itself across cell membranes: instead, water always follows the movement of salts or other solutes. In higher plants, the supply of water is almost totally generated by processes that take place in the leaves. As water evaporates from the leaves, osmotic pressure in leaf cells rises, attracting water by osmosis from the transporting xylem tissues in the plant's stem. We can think of this as a 'suction pressure' which is then transmitted down through the rigid stem cells to the roots, where it acts to pull in water from the soil. The suction pressure has to be sufficient to overcome any osmotic difference between the soil water and the water of the root cells, otherwise water will *not* move in, and the plant will wilt. In salt marshes, where external salt concentrations (and therefore osmotic pressures) are higher than those of the root cells, normal plants would wilt. But salt marsh plants overcome the problem by a mechanism in the leaves. The major difference between halophyte leaves and those of other plants is that the leaf cells of halophytes actively *absorb* salts instead of excluding them: their salt content is often greater than 600 mM NaCl. The raised osmotic pressure caused by the salt helps to draw water into the cells from the xylem, in effect increasing the suction pressure and drawing up water from the roots. From this it is apparent that while halophytes keep the salt concentration in the xylem lower than outside, it is important that *some* salt is allowed to leak in.

Halophytes also slow down the rate at which they lose water by decreasing transpiration, even though this may lead to problems of overheating. Leaf cuticles are thick and often waxy, and the stomata—openings to the outside—are placed in deep grooves to minimize water loss. Even so, halophytes are affected by lack of water: many species show declining growth rates as salinity rises, and these declines are probably attributable to lack of water rather than to a direct toxic effect of salt.

How to cope with waterlogging and anoxia

Besides the problem of high salt levels, salt marsh plants and mangroves have to deal with the fact that marshes and swamps trap very fine sediments with a high organic content and low permeability. They are therefore usually waterlogged and become not only anoxic, but contain many products of reduction such as sulphides, ethylene, and carbon dioxide, which are toxic to plant roots or inhibit their growth. How do the plants maintain their oxygen supply and deal with this toxic environment?

In salt marshes, plants avoid anoxia by two processes. First, their roots are concentrated very near the surface, where there is some oxygen. Second, they allow some oxygen to leak out from the root tissues, thus oxygenating the soil and so oxidizing the toxic chemicals produced by reduction. The salt marsh grass *Spartina anglica*, for instance, has a surface mat of fine roots and deeper anchor roots that have large, continuous air spaces within them. Surrounding the roots is an oxidized zone spreading out for 2–3 mm.

In mangroves, aeration mechanisms are more specialized. Several mangrove genera have aerial roots or pneumatophores (Fig. 5.4), protruding above the mud surface and containing a system of air-filled cavities that open to the outside through pores. *Avicennia*, for example, has pencil-shaped pneumatophores while *Bruguiera* has so-called 'knee roots' and others such as *Xylocarpus* have laterally flattened roots that run like ribbons above the mud. Even the prop roots of *Rhizophora* have extensive air-filled cavities and lenticels to allow gas exchange. In *Avicennia* at least, the pneumatophores contain enough air to maintain aerobic respiration while the swamp is covered by the tide. Not all mangroves produce pneumatophores, however, so it is evident that there are other mechanisms allowing mangroves to live in anoxic muds, of which we know nothing.

Salt marshes: the *Spartina* story

There are probably a few hundred species of plant worldwide that are tolerant of salt and typical of salt marshes. Of these, the species that cover most area are undoubtedly the cordgrasses, *Spartina* spp. (Fig. 5.3). Cordgrasses are originally an American and European phenomenon, spreading vigorously by underground rhizomes. *S. foliosa* is a native of the Californian coast, and in east-coast USA *S. alterniflora* and *S. patens* make up the majority of more than 2 million hectares of marsh. In Europe, the native *S. maritima* is presently rare, but was formerly common. From these original distributions, early settlers took *S. maritima* to South Africa, and more recently human influence has taken *S. alterniflora* to west-coast USA and to South America, where in some areas it replaces mangrove communities.

However, the major incident in this trans-world movement was the importation to southern Britain, in about 1829, of *S. alterniflora* from east-coast USA, probably in ships' ballast (Gray *et al.*, 1991). Here, at some time before 1870, it hybridized with the native British *S. maritima*, producing a sterile hybrid christened *S. townsendii*. This species was thought to be particularly vigorous, and plants from southern Britain were widely planted to reclaim and stabilize mudflats. In fact, *S. townsendii* itself is *not* very vigorous, but it gave rise to another species, *S. anglica*, by the process of chromosome doubling, some time before 1892. This species is fertile and extremely invasive, and has now spread, partly naturally and partly by planting, throughout Europe, to Tasmania, mainland Australia, New Zealand, and China. It has a high rate of growth, and because it tolerates submersion better than many other salt marsh plants, there have been fears that it might both reduce plant diversity on marshes and reduce mudflat areas sufficiently to interfere with factors such as the food of wading birds. Indeed, since the spread of *S. anglica* in Britain, species of waders such as the dunlin have declined, although there is no proof that decline of the area of available feeding grounds is responsible for this decline.

S. anglica is a young species, and is presumably still evolving. Its initial success story is being maintained in areas where it has been newly introduced. But in southern Britain, where it is now over 100 years old, some areas are suffering 'dieback': whole swathes of *Spartina* marsh die and the accumulated sediment may be eroded away. The cause of dieback is not yet understood—it is often associated with very wet, anoxic sediments in the centre of marshes, but may also occur at the seaward and landward fringes. The process of dieback has certainly reduced areas of marsh dramatically at some sites: between 1924 and 1981, the total area of marsh in Poole Harbour, UK, declined from 775 to 415 ha (Gray *et al.* in Jones, 1995). If waterlogging and anaerobic conditions are themselves caused by *Spartina* growth, as some observers believe, the species is in effect limiting its own distribution. Presumably some equilibrium will finally be reached, as dieback is a feature of the well established populations of *S. alterniflora* in eastern USA.

A superficial life: salt marsh and mangrove snails

Despite the protection of a close-knit canopy of plants, physical conditions on the surface of salt marshes and mangrove swamps can be harsh. Salinity may rise and fall drastically. Periodic—and at the upper levels of marshes, unpredictable—cover by the tides prevents air-breathing and may sweep away animals that have no appropriate attachment mechanisms. Mobile animals in these environments have to opt between two basic lifestyles. Some burrow into the sediment. Others stay on the mud surface or live on the stems and leaves of plants. In other words, their lifestyles are equivalent to those of infauna and epifauna on unvegetated flats. Here we deal with some epifaunal species, the snails.

Physiological tolerance of extreme changes in salinity and water supply is a major characteristic of many salt marsh snails (Little, 1990). The European pulmonate *Ovatella myosotis* can survive on mud with salinities from fresh water to nearly three times that of sea water. On the Atlantic coast of the USA, another pulmonate, *Melampus* (*Leucophytia*) *bidentatus* can lose up to 80% of its body water and still survive.

Ability to breathe air is another essential, because, as we have seen, most marshes and swamps grow above MHWN. Hours of submergence at any one height will vary at different sites depending upon the tidal regime, but we can gain some perspective by considering figures for Poole Harbour, UK. Here, marsh at MHWN will be covered by more than 700 tides in a year, totalling 3600 h of submergence. At MHWS, only 50 tides cover the marsh, giving 400 h submergence, or less than 5% of the year. Pulmonate snails like *Ovatella* and *Melampus* have a mantle cavity without gills, which is adapted as a lung. But how do snails of marine origin, such as the winkles, survive? *Littoraria* spp. breathe air. They do so using gills, but the gill filaments are reduced to small triangular leaflets on the left side of the mantle cavity which extend as folds across the cavity's roof (Reid, 1986). They are presumably rigid enough to provide increased surface area in air without collapsing, as the larger gills adapted to function in water would do. *Littoraria irrorata*, a common species in east-coast USA, actually retreats upwards on *Spartina* stems as the water rises, so is seldom covered by the tide. When it is covered, it may become bloated because the incoming sea water often has a low salinity. It is also subject to devastating predation from the blue crab, *Callinectes sapidus*, as tethering experiments on the marshes have shown. Snails tethered on the marsh surface are five times more likely to die than those tethered high up on *Spartina*. Other species of *Littoraria* are common on mangrove trunks and leaves throughout the tropics. All of them avoid water cover, some even showing tidal migrations up and down, moving upwards just in front of the tidal rise.

Another problem connected with living in marshes or swamps seldom covered by the tide is that of reproduction. Many marine animals liberate their eggs and sperm—or larvae—into the sea so that the larval stages can spend some time in the plankton. This approach has its problems at high tide levels, and some snails, such as *Ovatella*, lay egg masses which hatch as miniature snails with no pelagic larval phase. *Melampus*, however, still has planktonic larvae, the veligers. Their successful development is made possible because the reproductive cycle is geared to that of the tides (Fig. 5.6). The snails copulate and lay eggs when high spring tides cover the marsh and dampen the mud. Larvae hatch from these eggs after about 2 weeks, coinciding with the next set of high spring tides so they can disperse into the plankton. Planktonic life is short, and after a further 2 weeks the larvae are ready to settle back on the marsh on the third successive set of spring tides.

How the *Littoraria* species in mangrove swamps deal with the problem of getting their larvae to sea and back is not known. Some produce pelagic egg masses, while others brood their young to the stage of veliger larvae, so all are presumably released when high tides cover the swamps.

Fig. 5.6 Reproductive cycle of the snail *Melampus bidentatus* in east-coast USA. Black bars show when spring tides cover the marsh. Reproductive phases (hatched blocks) are geared to these periods. Further phases producing later cohorts of young (not shown) are geared to later tides, while juvenile growth (circles and line) occurs in a smooth curve despite the irregular cover by tides. (After Russell-Hunter *et al.*, 1972.)

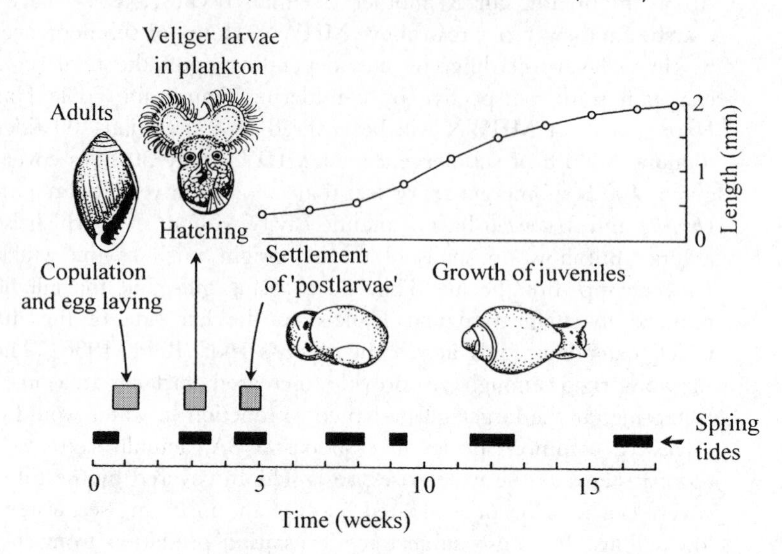

Life in burrows: salt marsh beetles

Burrowing into the sediment provides protection against surface predators, and is some defence against being washed away. Burrows also ameliorate the effects of rapid changes in salinity and temperature that occur on the salt marsh surface. For air-breathers, however, burrows do create some disadvantages: the water can flood them at high tide, and the tide can also block the entrance with debris, suffocating the inhabitants. Nevertheless, many beetles adopt a burrowing strategy, and the staphylinid *Bledius spectabilis*—recognizable from its striking red wing-cases—frequently has large populations at the seaward edge of European marshes. *Bledius* not only lives in the burrows, the female lays her eggs there and the young hatch and grow up in the burrow before dispersing.

The burrow lifestyle is only possible for *Bledius* because, unlike most other invertebrates, the female beetle shows parental care (Wyatt, 1986). She digs a burrow with a narrow neck, rather like a wine bottle, and lays her eggs in small side-chambers (Fig. 5.7). The 2 mm opening delays entry of the tidal water but only temporarily, so when the tide rises the female blocks the opening with mud, unsealing it as the tide falls. Her behaviour is a direct reaction to water cover—no tidal rhythm is involved, probably because tidal cover over salt marshes can be quite variable.

The female *Bledius* also defends the burrow against predators, such as the carabid beetle, *Dicheirotrichus gustavi*. This predator has no burrow and retreats to cracks in the mud during the day. It has an endogenous circadian rhythm

Fig. 5.7 Burrow of the beetle *Bledius spectabilis*, showing eggs in individual chambers, and developing larvae. The predatory beetle *Dicheirotrichus gustavi* eats *Bledius* larvae. (After Wyatt, 1986, and photographs by C. Little.)

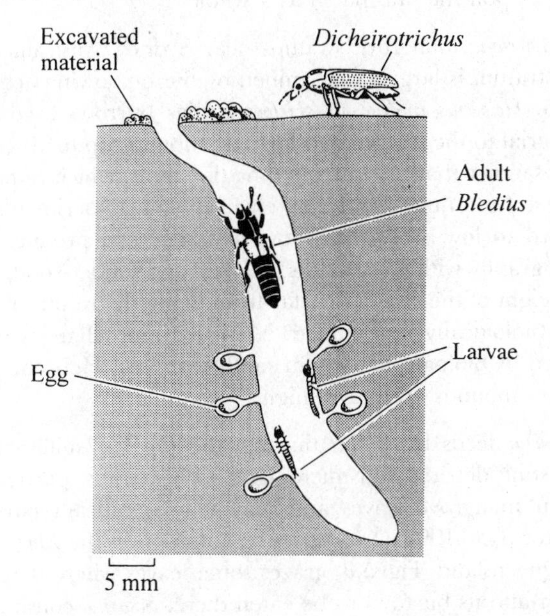

which triggers emergence at night, and it then preys on *Bledius* eggs and larvae. When female *Bledius* are experimentally removed from burrows, *Dicheirotrichus* may eat the entire brood and then take over the burrow. Even if *Dicheirotrichus* does not take over, however, the burrows become blocked by sediment and the eggs often become infected by fungus, so survival is low. Nor is *Bledius* alone in its habits: several other species of salt marsh beetles look after their brood to some degree, showing that parental care by the female is essential in a habitat that can be regarded as harsh—at least for animals of terrestrial origin.

Life in burrows: crabs and their effects

Burrowing crabs are a feature of many mangrove swamps and salt marshes. Population densities of *Uca pugnax* in east-coast American salt marshes, and of *Heloecius cordiformis* in Australian mangrove swamps can both be greater than $50/m^2$. We have already seen that crab burrows can increase water flow through mangrove swamps (p. 23). What overall effects do crabs have?

In a New England salt marsh, experiments show that the fiddler crab *Uca pugnax* actively stimulates growth and production of the cordgrass *Spartina alterniflora*. When the crabs are removed from areas where the *Spartina* forms tall swards, cordgrass production of shoots and leaves declines by nearly 50% (Bertness, 1985). This effect almost certainly occurs because the crabs normally improve drainage through the soil. Good drainage counteracts the effects produced by *Spartina* itself—the accumulation of organic detritus leading to reduced drainage, and anoxic soils. In areas where *Uca* is uncommon, *Spartina* forms only a short sward and has low rates of production, so the crabs have an enormous effect upon the salt marsh as a whole.

In *Avicennia* mangrove swamps near Sydney, Australia, the topography of the substratum is largely determined by the burrowing activities of the semaphore crab, *Heloecius cordiformis*. *Heloecius* digs burrows as deep as 40 cm, bringing material to the surface and forming mounds up to 10 cm high. These are drier and sandier than the surrounding flat areas, which remain wet even at low tide and so form quite a different environment. Experiments in which the crabs are added to low, wet areas show that they can produce substantial changes in topography within 4 months (Warren and Underwood, 1986); so the variations in height of the swamp are far from being the results of physical influences, but are biologically determined. More extreme changes in topography are produced by biological action in south-east Asia. Here the mud lobster, *Thalassina*, makes mounds up to 2 m high.

Heloecius feeds by sifting through the top few millimetres of the substratum, ingesting detritus and microalgae. Other crabs, particularly grapsids, feed on fallen mangrove leaves and may play significant parts in recycling organic matter (Lee, 1998). *Sesarma messa*, common in the *Rhizophora* mangrove swamps of Queensland (Fig. 5.8) grazes some leaves where they fall, but pulls many of them into its burrows to be eaten there. *Sesarma* consumes an astonishing 30% of the total leaf fall over the year in some swamps, thus retaining organic material in the swamp that would otherwise be removed by the tide (Robertson, 1986). *Sesarma* burrows are also important for the trees: they perform the same functions as those of *Uca* on American salt marshes, so aiding growth and

Fig. 5.8 The effect of the crab *Sesarma messa* on leaves of the mangrove *Rhizophora stylosa* in a mangrove swamp in Queensland, Australia. Yellowing leaves were tethered individually or placed in nylon mesh bags to prevent access by the crabs. *Sesarma* removed the majority of the leaves to which it had access within 2 weeks. Bars show 95% confidence limits. (After Robertson, 1986.)

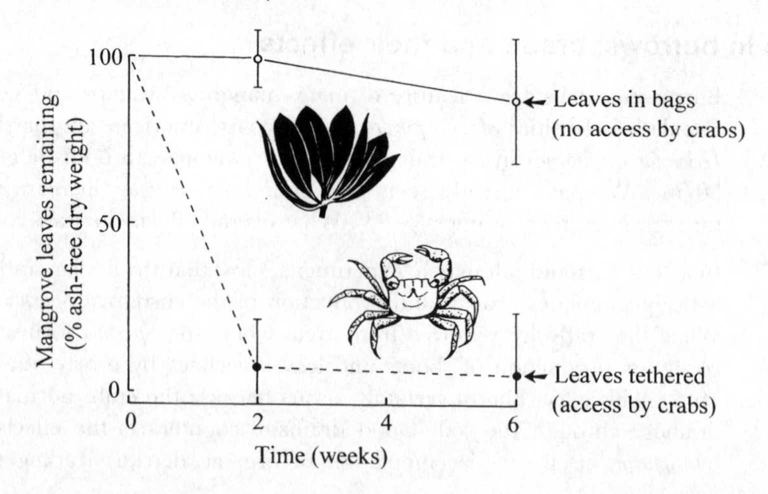

production of the *Rhizophora*, especially along the creek banks (Smith *et al.*, 1991).

These three examples demonstrate how important crabs can be in determining the structure and function of mangrove and salt marsh ecosystems. They also emphasize that the biology of individual species can rarely be viewed in isolation from others, and we now turn to a consideration of biological interactions and the organization of food webs.

Biological interactions and food webs

Interactions between plants: what causes plant zonation?

On most salt marshes, plant species are clearly distributed in bands parallel to the shore. In east-coast America, for instance, *Spartina alterniflora* occupies the lower marsh while it is replaced by *Spartina patens* higher up. In Europe, *Salicornia* or *Spartina anglica* are found on the lower marsh, but are replaced by the grass *Puccinellia* and many broadleaved plants at higher levels (Figs 5.2 and 5.3). In mangrove swamps, distributions may be harder to follow, and in large swamps the trees form a patchwork or mosaic. But even here, species show a clear relation to tidal height: in America, *Rhizophora* is found at the seaward edge, *Avicennia* and others towards the land. In the Indo-Pacific, these two genera show a reverse pattern. What causes these 'zonation' patterns?

We have already discussed zonation on sandy shores (p. 46). Here, many authorities conclude that physical factors such as drainage, particle size, and salinity are important in controlling animal distribution, but that biological interactions such as competition can modify these controls. On angiosperm-dominated shores, the 'classic' theory has been that plants show a 'succession' over time, which is reflected in their spatial distribution. Thus plants low on the shore trap sediment, and as the marsh rises conditions become more favourable for other species, which can then colonize. The original species then disappear, either because they have actually created conditions which are suboptimal for themselves, or because of competition with incoming species. This picture portrays marshes and swamps as continually growing upwards from MHWN, and finally forming part of the terrestrial environment. Higher levels thus represent later phases in the succession.

The situation is, however, more complicated than this. First, we should note that the zonation along a gradient of tidal height (a spatial feature) has not always been produced by a straightforward linear succession (a temporal feature). For example, the advent of *Spartina anglica* in Europe has produced salt marshes in which the lower parts are dominated by this grass, while higher levels have other species such as *Puccinellia*. But *Spartina* came *later* than *Puccinellia*, so is not part of a classical succession. Second, many marshes show a mosaic pattern of vegetation which appears to be stable rather than evolving towards a later stage in a succession, and it now seems that marshes often remain stable until some external factor triggers an increase in accretion (or erosion) rate. A third point, well demonstrated in mangrove swamps, is that although there may be a recognizable change in predominance of different species from the seaward to the landward edge of the swamp, individuals of some species can be found at almost all heights. In east Africa, for instance, *Avicennia marina* may be abundant at the top of the swamp as well as at the seaward edge, but is scarce in the centre (Macnae and Kalk, 1962). Instead of thinking in terms of successions, it may therefore be more useful to consider vertical distributions in terms of the various factors that determine these distributions at the present time.

There is some agreement that many angiosperms on the shore are limited in their downward spread by physical factors related to tidal submersion, while their upward spread is controlled more by competition with other species. Note that this is precisely the opposite of the consensus about rocky shores, where animals and algae of marine origin are thought to be limited by competitive factors at their lower limits and by factors related to time out of water at their upper limits (Little and Kitching, 1996). Is the situation in salt marshes and mangrove swamps really this simple?

The major physical factors that limit the spread of angiosperms downshore are usually thought to be salinity—both its absolute value and its degree of fluctuation—and a set of effects related to the number of submergences experienced, or perhaps the total submersion time. Different plants have different tolerances of salinity which may act to control distribution directly, but little is known of how these control mechanisms actually work. One possibility is that salinity affects the physiology of the root cells, as these are directly exposed to saline water rather more than the leaves and stems. However, many halophytes are tolerant of a very wide range of salinity, and differences in distribution are more likely to be governed by competitive effects which in turn are controlled by a wide variety of factors. Salinity certainly poses indirect problems for the plants too: saline soils reduce nitrogen availability, so plants in high salinity may effectively lack nutrients. In general terms, salinity in the sediment increases downshore, because it is not diluted by land drainage, and it is obviously a factor that must be taken into account when trying to explain distribution.

Submersion time affects a number of variables important for plant growth: further downshore, the time for photosynthesis is reduced, and the tide may deposit more sediment on the leaves, thereby reducing photosynthesis even more. The tide also creates mechanical scour, and particularly for seedlings this may be critical—could this explain why mangroves generally have large propagules? As tides alternately cover and uncover the plants, temperature fluctuates considerably, and this may be an additional stress on the plants' physiology. One of the major problems in sorting out the effects of these physical factors is that they all change together downshore, making critical experiments very difficult.

At the top of the shore, competition between salt marsh plants is well shown by experiments carried out in the Netherlands. Here the top of the *Spartina anglica* zone gives way to *Puccinellia maritima*. In adjacent plots differing in height by only 4 cm, experiments in which one of these two species is removed have dramatically different effects (Gray in Allen and Pye, 1992). In the lower plots, removal of *Spartina* is followed by increased growth of *Puccinellia*, while the removal of *Puccinellia* has no effect on *Spartina*. In the upper plots, the reverse is found: removal of *Puccinellia* is followed by increased growth of *Spartina*. In the lower plots, therefore, *Spartina* normally outcompetes *Puccinellia*, whereas 4 cm higher up *Puccinellia* wins the competition. The balance of competition here is presumably determined by the underlying physical conditions which change with height on the shore, as we have just discussed.

Evidently the suggestion made earlier, that competition limits the upward spread of species while physical factors limit their downward spread is far too simple. Not only does competition vary under different physical conditions, but the plants themselves actually alter those physical conditions by producing organic detritus, slowing water currents and so on. In effect there is a complex system with what we might term 'feedback' loops, in which any changes to the system alter conditions in which plants interact. Until salt marshes and mangrove swamps are regarded in this light, and not in terms of the effects of one variable such as salinity, it is unlikely that any true understanding of the causes of plant distribution will be reached.

Animal zonation

As on mudflats and sandy shores, animals in salt marshes and mangrove swamps also show patterns of zonation. In mangrove swamps particularly, these patterns are complicated by the three-dimensional nature of the trees themselves: some animals are restricted to the mud surface, others colonize the roots and trunks of the trees, while more terrestrial species live in the canopy. Distinct vertical 'zones' can be recognized within a swamp, but these cannot be related strictly to tidal height: the *Nerita* zone in a Malaysian swamp, for instance, rises in height from the seaward edge to the rear of the swamp (Fig. 5.9). In addition, some species are found only towards the seaward edge. Bivalves such as the oysters *Isognomon* are found in a narrow band at the seaward fringe, and burrowing polychaetes do not stretch far into the swamp. These distributions may be explained by the diminishing number of larvae that reach into the interior of the swamp, or by physical conditions that change with decreased water flow away from the seaward edge.

Animals living on or in the mud probably differ significantly from those living on the trees in the factors that determine distribution. For instance, the benthic crabs in a Jamaican mangrove forest show patterns of zonation that clearly relate to tidal levels, and the consequent physical properties of the mud (Warner, 1969). In contrast, arboreal winkles, *Littoraria*, found in Indo-Pacific swamps, show both *vertical* zonation (up the trunks of the trees) and characteristic *horizontal* patterns of distribution that are not related to tidal height (Reid, 1985). Thus some species are common at the seaward edge, others in the

Fig. 5.9 Diagrammatic section through a Malaysian mangrove swamp, showing the distribution of five faunal zones: *Littoraria* zone (left-handed hatching), *Nerita* zone (right-handed hatching), *Uca* zone (vertical hatching), bivalve zone (horizontal hatching), and burrower zone (thick diagonal hatching). Also shown are some of the other dominant fauna. Heights are given above chart datum, which is the level of the lowest tides. (After Berry, 1964.)

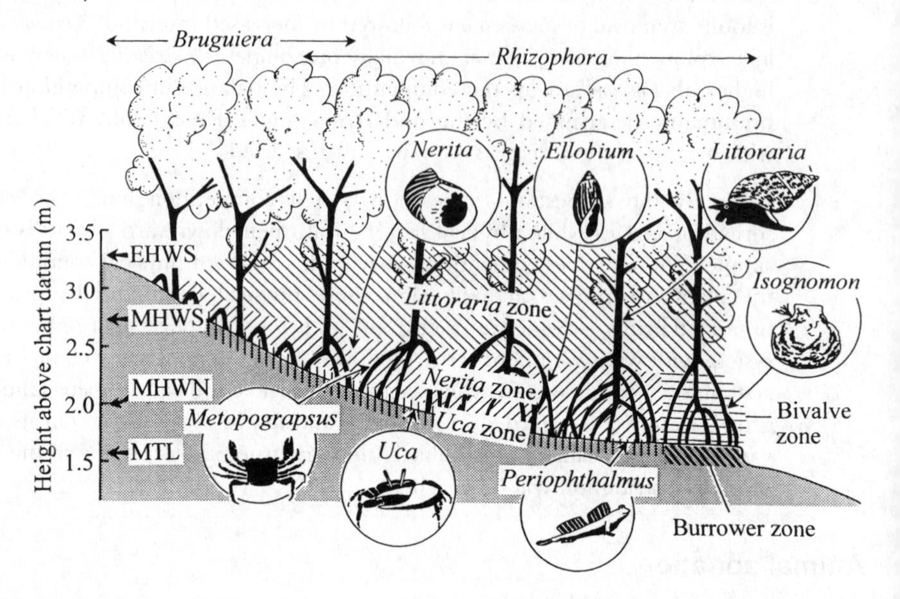

middle, and still others near the landward edge. Some of these differences in distribution may be related to the distribution of predatory arboreal crabs, or to distribution of larval settlement. As with zonation on other types of shore, there is no simple answer to the question of what determines distribution.

Primary production and food webs

Salt marsh plants and mangrove trees are highly productive organisms, fixing large amounts of carbon dioxide and converting it into organic molecules. Consequently they provide enough food for large numbers of animals, and both marshes and swamps support high animal biomasses. Measurements of primary productivity in these semi-aquatic environments are not, however, at all easy, and there are in fact relatively few reliable figures, so we begin with some cautionary points.

Net primary production is the total amount of energy—or mass—incorporated by photosynthesis, less that which is lost in respiration. It can thus in principle be measured by recording carbon dioxide fluxes, but this is a technically challenging process, especially during tidal cover. Alternatively, and more widely, production can be estimated if all the gains and losses in plant biomass are measured over a given period. This method is also challenging, however, as the losses may involve a wide range of routes: some biomass is eaten by herbivores, dead particulates are lost as litter, the soluble fractions leach out at high tide, and the roots may also lose material as exudates. The method therefore involves a good deal of approximation. Many observers have chosen to measure only above-ground production (i.e. by the leaves and shoots), while others include the roots. Some observers have worked only in the maximum growing season, and have to make approximations to estimate production and loss over the entire year. There is also, confusingly, a plethora of recording units, from units of energy such as kilojoules and kilocalories to units of mass such as total dry weight or weight of carbon. With these problems in mind, it is evident that production estimates will be, indeed, only estimates.

Here we will concentrate mainly upon salt marshes, keeping strictly to measurements in terms of weight of carbon. The highest producers are salt marshes dominated by the tall form of *Spartina alterniflora*, some of which produce over 2000 gC/m^2 per year above ground. Added to this is the below-ground production, which may be about the same or more than the above-ground figure. Algae on the mud surface add about 20% to the above-ground *Spartina* figure, giving a total for the marsh of possibly as much as 5000 gC/m^2 per year. This is an extreme value, however, and for many salt marshes production figures are more of the order of 700–1000 gC/m^2 per year. Tropical mangroves may reach 1800 gC/m^2 per year, but an average value for northern Australia and Papua New Guinea is only 950 gC/m^2 per year (Robertson, 1986). Part of the variation in productivity is related to latitude—highest rates for salt marshes are found in Georgia, USA, while rates decline further north on the Atlantic coast of America. How do these figures compare with other ecosystems? Tropical seagrass beds may reach as much as 7000 gC/m^2 per year, but tropical rain forest, coral reefs, and the best cultivated land reach only about 1800 gC/m^2 per year. Rates on some salt marshes are thus high, but the average salt marsh is more in line with rates in temperate forests and grasslands.

What happens to the high biomass that results from this productivity? We start by considering observations at Sapelo Island, a salt marsh in Georgia, made by Teal (1962) and considerably extended by later studies (Pomeroy and Wiegert, 1981). This classic example has strongly influenced thoughts about energy flow on salt marshes and provides a good introduction to salt marsh food webs. Two surprising points arise from this study (Fig. 5.10). First, very little of the production by *Spartina* or algae is consumed directly by herbivores—less than 4%. The rest is degraded to detritus, and passes along the detritus food chain. Second, a large proportion of this detritus is calculated not to be utilized on the marsh. Early estimates suggested that 45% of it is exported to the neighbouring estuaries and flats, while later studies revised this figure to nearer 35%. We shall consider this second conclusion in the next section, and for the moment follow the links in the food web on the marsh itself.

Although herbivores do not process much of the *Spartina* biomass, some hemipteran insects such as *Prokelisia* can be abundant, sucking out the plant juices. They and other insects provide a food source for the abundant spiders. As most plant material is degraded to detritus, detritivores understandably

Fig. 5.10 Energy flow in a Georgia (USA) salt marsh. Units are given in gC/m^2 per year. To simplify the diagram, respiratory losses are not shown. Circles and black arrows show estimates after Teal (1962). Rectangles and white arrows show some suggested additions for salt marshes in this and other regions. DOM, dissolved organic matter.

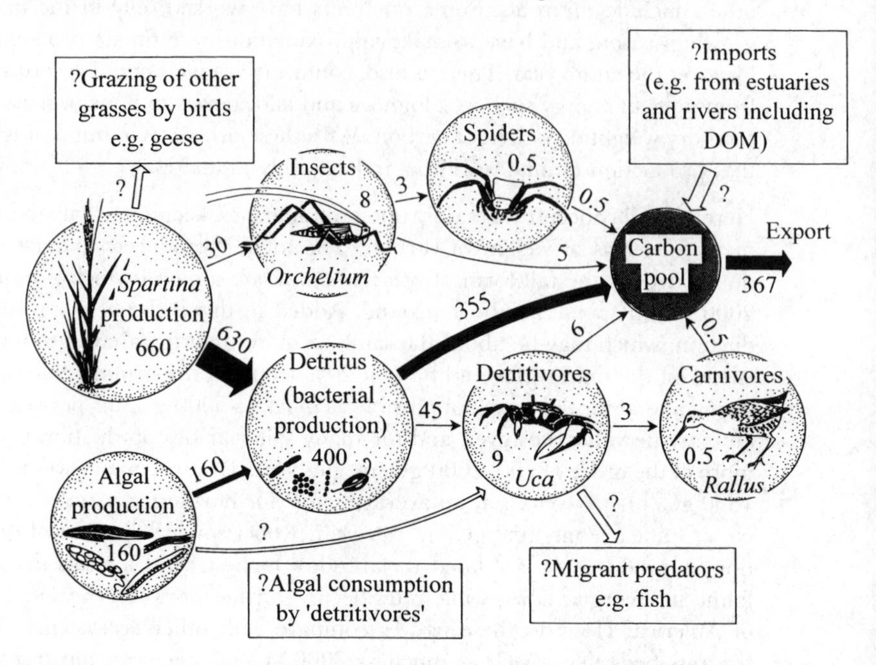

provide the biggest fraction of consumer biomass. Fiddler crabs, *Uca*, and grapsid crabs such as *Sesarma* eat detritus, and as suggested on the figure, they also eat microalgae. Annelids, nematodes, and snails (such as *Melampus*) all process significant quantities of the detritus that forms on the marsh surface. Bivalve molluscs such as the mussel *Modiolus* feed on particles of detritus suspended in the water. These detritivores provide food for a second set of carnivores, the resident mud crabs, *Eurytium*, and visiting racoons, *Procyon lotor*.

How typical is this food web in Georgia? Later estimates on a wide range of marshes agree that direct consumption of angiosperms by herbivores usually amounts to less than 10% of production, despite the fact that a wide variety of insects attacks the plants, from aphids to grasshoppers. Indeed, insect herbivory may be very low on some marshes. In Britain, insects have been estimated to consume only 0.3% of the net annual above-ground production of *Spartina anglica* (Gray *et al.*, 1991). The food webs are thus all thought to be primarily detritus-based. However, studies using stable carbon and nitrogen isotopes (see p. 89) have shown that 'detritivores' often eat benthic microalgae and allochthonous detritus that originated not on the marsh but on dry land. For instance, in a marsh in Southern California (Page, 1997) where the major vascular plant is *Salicornia*, values of δ^{13}C in amphipods and snails on the salt marsh surface are about −20‰, compared with −26 to −29 for *Salicornia* and

−14 to −21 for benthic microalgae. These animals therefore probably eat *Salicornia* and some algae. But invertebrates from the salt marsh creeks have values more in the region of −16 to −15, suggesting almost total dependence upon microalgae. Studies on a marsh in Brittany, France, show that many of the invertebrates on the salt marsh surface there depend upon benthic diatoms (Créach *et al.*, 1997). On a marsh in east-coast USA, the mussel *Geukensia* has $\delta^{13}C$ values as extreme as −27‰, suggesting this species feeds mainly on detritus of terrestrial origin, brought down by rivers. The importance of algae and angiosperms in the food web is evidently variable and needs further study.

The Georgia salt marsh also has its own geographical bias. Many marshes in Europe and Australasia are not so dominated by crabs. For example, in one marsh in Britain (Fig. 5.11), smaller crustaceans and snails are the main detritivores, together with beetles. In this marsh the widest diversity of predators is provided by the insects and spiders, although crabs like *Carcinus* do invade the marsh at high tide. Marshes in America were formerly used as haymeadows, and many marshes throughout the world are still used for grazing by sheep and cattle. Grazing can remove most of the above-ground biomass, and may also influence plant communities: sheep grazing can convert *Spartina* marsh to the grass *Puccinellia*, for instance (Ranwell, 1972). Grazing by other vertebrates varies widely over different geographical areas. Wintering snow geese may arrive in thousands on east-coast USA, and because they feed mostly on roots and rhizomes, they may uproot even more of the angiosperm biomass than they consume. Grazing by these geese may also affect species composition, promoting dominance of *Puccinellia*. In contrast, feeding by Brent geese on a marsh in eastern Britain leads to decline of several species, and in cages *excluding* the geese, *Puccinellia* increased in abundance (Rowcliffe *et al.*, 1998).

Fig. 5.11 Distribution and trophic levels of some common invertebrates on the lower regions of salt marshes in the Severn estuary, UK. (After Little, 1992.)

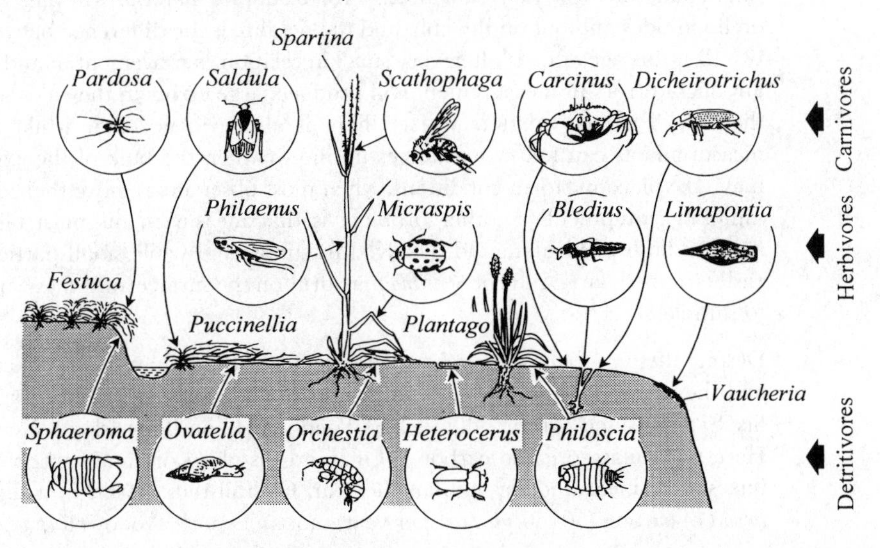

Another factor missing from Teal's study in Georgia is the impact of species that visit the marsh at high tide. In some marshes, fish have a significant predatory effect but this may vary over the marsh. In the Great Sippewissett Marsh, Massachusetts, for instance, the Killifishes, *Fundulus* spp., prey on the infauna of sandy-organic sediments but have little effect on muddy areas (Sarda *et al.*, 1998). It is therefore likely that food webs vary not only between but within marshes.

One of the main conclusions of recent research is that it is unwise to generalize from one marsh to another. Detritus from different species of plant is utilized to differing extents, and the effects of different grazers can vary enormously. As we shall see in the next section, the balance between imports to the marsh and exports from it also changes with geographical location. Nevertheless, the calculated food web for the Sapelo Island system remains extremely useful as a basis for viewing the factors that vary from one marsh to another.

Exchange between marshes/swamps and local estuaries

The Sapelo Island marsh is thought to export between one-third and one-half of its primary production to local estuaries, and this calculation has led to a widespread view that salt marshes are of major importance in providing a detrital food source for estuarine organisms. But is this large amount of export typical? Does it apply to mangrove swamps?

First it is important to realize that making estimates of the overall exchange between marshes and nearby estuaries is extraordinarily difficult. The most widely used technique is to take water samples from a creek that drains a specific area of marsh, and to analyse these for carbon content. But exchanges must be summed over an entire year (at least) to make sense, and during this time conditions will vary enormously. For example, material will pass inwards on flood tides and out on the ebb, and the net flux is the difference between the two. But this net value is often very small in relation to movement in and out, so any slight errors in measurement will produce large errors in the calculation of the flux. To make matters worse, there is also evidence that while routine measurements can follow exchanges in fine weather, the bulk of the exchange may take place in storm conditions, when most observers remove their delicate sampling equipment! Another problem is that measurements must take into account both particulate and dissolved fractions, and while small particles are easily trapped, large rafts of *Spartina* may drift on the surface, and these are hard to sample.

Despite these problems, though, several detailed studies have been undertaken, and they show a broad spectrum of results. At one extreme are the exporters, like Sapelo Island. At the other is Flax Pond marsh, on Long Island, New York. Here particulate organic carbon (POC) and dissolved organic carbon (DOC) fluxes were measured throughout the year. Overall these result in a slight net *import* of carbon (50–60 gC/m^2 per year), not an export (Woodwell *et al.*, 1977).

This overall figure hides the variation over the year: there is a net loss of DOC in summer and a net accumulation in winter, while POC shows a net input throughout the year. One extra complication is the possibility that carbon might also be exchanged as living biomass: because fish use the marsh as a nursery ground and then when they have grown move offshore, their increase in weight represents an export. In Flax Pond this is calculated as about 20 gC/m^2 per year.

Between the importer Flax Pond and the exporter Sapelo Island comes a range of degrees of export. The consensus at present, however, is that while many marshes export detritus, this does not usually provide a very large fraction of the detritus available to organisms in local estuaries. This view is supported by the use of stable isotopes of carbon and nitrogen in examining estuarine faunas. As seen previously (p. 112), these suggest that the majority of estuarine detritivores eat benthic algae and detritus derived from phytoplankton and from allochthonous sources such as rivers to a greater extent than detritus exported from salt marshes.

However, analyses of the fluxes of elements other than carbon give rather different pictures. In particular, observers have been interested in nitrogen budgets, as plant growth may be controlled to some degree by nitrogen supply. Here again, marshes differ widely. One marsh in Essex, Britain, is roughly in nitrogen balance. Sapelo Island imports nitrogen, while Great Sippewissett exports it. These studies show that marshes may be very important in affecting conditions in local estuaries, even if they do not export large amounts of carbon.

The situation with regard to mangrove swamps is, if anything, less clear because there have been fewer detailed studies, especially in the tropics. In Florida, mangrove detritus is mostly flushed out of the forests by the movement of tidal currents. Leaves, flowers, and other products are then grazed by crabs, snails, and amphipods and subjected to bacterial decay in local shallow water. In Australia, 30%—and possibly more—of the leaf fall is removed by crabs (see p. 106) and either eaten or buried, so that a reduced proportion of primary production is available for export (Robertson, 1986). Nevertheless, most mangroves are probably net exporters to some degree, and fuel offshore food webs. Like salt marshes, too, they often act as nursery areas, and offshore fisheries for shrimp in the tropics may depend upon the integrity of local mangrove swamps.

An investigation of exchange between a mangrove swamp in southwest India and the local estuary details some of the ways in which this exchange can be important (Wafar *et al.*, 1997). Here the average litter fall in the mangrove swamp produces approximately 500 mg/m^2 per day of POC and nearly 900 mg/m^2 per day of DOC. This compares with a total production by phytoplankton in the estuary of about 600 mgC/m^2 per day; i.e. the mangrove swamp is roughly twice as productive as the estuarine phytoplankton. However, the production by mangroves occurs in an area of only 1600 ha, and when exported this is spread over an area of estuary of about 9200 ha; i.e. it is effectively diluted. When this dilution is taken into account, and allowance is made for carbon lost from the estuary by tidal flushing, the mangroves are calculated to supply only 80 mg POC and 140 mg DOC/m^2 per day. Over the year this provides an average of 37% of the carbon available in the estuary. Naturally this supply varies over the course of the year: during most of the year it provides much less than half the total, while phytoplankton input is high. But during the monsoons, despite the fact that the fall of leaves is then minimal, they provide more than the phytoplankton which is also at its minimum (Fig. 5.12).

Detritus exported from this Indian mangrove swamp has two other effects on the estuary. First, the DOC stimulates microbial growth in the estuary, and so fuels the microbial food chain, essentially providing more food for the detritivores. Second, the dissolved nitrogen that is also exported stimulates the growth of phytoplankton—which in turn provides more detritus for the benthos. As with salt marshes, mangrove swamps can thus affect local estuarine conditions to a greater extent than consideration of carbon budgets alone would imply.

Techniques

Because salt marshes and mangrove swamps are dominated by angiosperms, their study requires some specialized techniques, and we have already discussed methods for measuring primary productivity, for estimating the flux of material into and out of marshes, and the use of stable isotopes in determining the origins of detritus and of animals' food. Many general approaches can, however, be adapted from those used on mudflats and sandy shores. Baker and Wolff (1987) give a good introduction to survey methods for salt marshes, and Snedaker and Snedaker (1984) review methods for mangrove swamps.

Fig. 5.12 The relative importance of phytoplankton (filled circles) and mangrove detritus (open circles) in the supply of organic carbon to an estuary in southwest India. Carbon flux is calculated as the amount supplying 1 m^2 of estuarine water. (After Wafar *et al.*, 1997.)

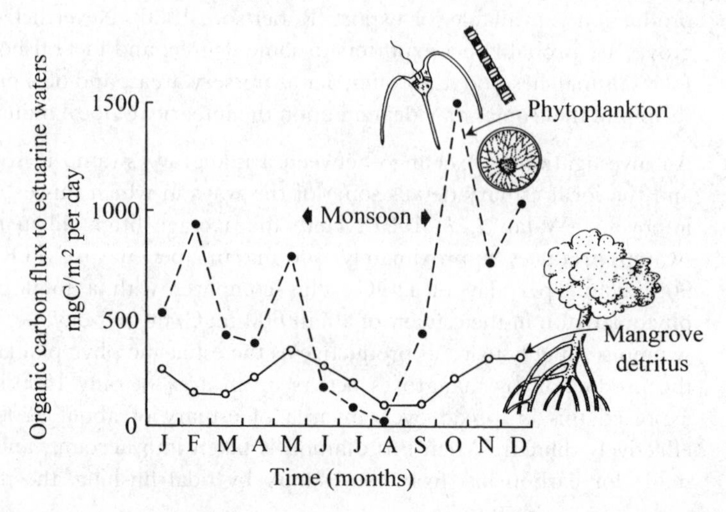

Access

It is usually much easier to work on salt marshes than on mudflats, although the *Spartina alterniflora* marshes of eastern America present special problems because they can grow to a height of 3 m. Generally the substratum is firm, and because most marshes are above MHWN, coverage by the tide is seldom prolonged. In addition, salt marshes allow easy access to salt pans—microcosms that may help in understanding some aspects of mudflat ecology. Many of the organisms that live in salt pans are typical of mudflats—in Europe they contain animals such as *Arenicola*, *Nereis*, *Corophium*, and *Hydrobia*, for example. The pans are like miniature aquaria, allowing direct observations of the activities of their inhabitants (Steers, 1960).

Mangrove swamps, in contrast, usually provide no easy access, and in those that do—the *Avicennia* type—they are fragile and easily damaged by trampling. Recent developments that allow observation without causing excessive damage have been the construction of walkways to provide access to the swamp surface, and towers to provide access to the tree canopy.

Describing the distribution of vegetation

Quadrats of various sizes provide the most commonly used basis for recording plant distribution. The most accurate method is called 'point sampling', in which vertical pins make contact with a set number of points in each quadrat, or crosswires allow viewing from above to give the same effect. Recording the number of points contacted for each species allows calculation of values such as percentage cover (Baker and Wolff, 1987).

But the subsequent analysis of data collected in this way has led to many divergent views about whether plants can be regarded as living in 'communities' or 'associations', and how these should be described. There are not usually many different species in one area of salt marsh, and these tend to occur in a relatively limited number of species groupings which are usually called 'associations'. This means that various areas of marsh can be described in terms of the distribution of these associations. Many have conspicuous dominant species, and these dominants are used to provide names for the associations. For example, an area dominated by the sea aster, *Aster tripolium*, is called an Asteretum, while one dominated by *Spartina* is a Spartinetum. The definition of such associations began as a series of subjective judgements, but to a great extent these have now been replaced by one of two kinds of more objective classification schemes (Adam, 1990). Phytosociological schemes separate out associations on the basis of varying proportions of species, and may name each association after a dominant or series of dominants. In contrast, numerical classifications calculate the similarity between quadrats in sampled areas depending upon characters such as species held in common, abundances of each species, species unique to the quadrat, and so on.

There are, however, problems with either approach. One such problem is that the vegetation often occurs in mosaic patterns, so quadrats often sample more than one association. Another is that concentration upon dominants tends to obscure variations in the abundance of less dominant species: an Asteretum could have a wide variety of other less abundant species, depending upon the site, unless the concept of an Asteretum is rigidly defined. Indeed, it is important to realize that the concept of an association is an abstract one, and there is always a danger that data which do not really fit the concept are squeezed into it! We should also emphasize that salt marshes frequently have an angiosperm layer well above the mud surface, but an algal layer on the mud surface itself—and associations of algae do not always coincide with those of angiosperms. What happens if there is no dominant species? As we have noted earlier (p. 95), some salt marshes have been termed a 'general salt marsh community' because they contain several equally common species. Unfortunately, the term has now been used for several quite different associations, so is not a particularly helpful one.

Nevertheless, recorders who wish to compare salt marshes in different areas need some classification scheme. In Britain, for example, there is a National Vegetation Classification (NVC) which identifies 'communities' characterized by dominants which are given a name and number. This is certainly helpful for descriptive scientists but cannot be said to stimulate an investigation into the factors that influence variations in plant distribution. Examples of the NVC are given by Davidson *et al.* (1991).

6 Life at the bottom: sublittoral sediments and community structure

Below the level of low tide, species common in the intertidal zone are replaced by others. In part, the changes are related to increasing depth, so there are species-characteristic depth zones. In some cases the reasons for this sublittoral zonation are fairly easily understood—for instance algae and seagrasses do not penetrate to great depths because they need light for photosynthesis. But in general, we know little of why species are distributed in this way, and there have been no rigorous tests to examine why intertidal species are not found in the sublittoral. Is this to do with physical tolerances, or is it more likely to be explained by changing balances of competition and other biological interactions?

When distributions of benthic fauna are examined over wide ranges of depth and latitude, further unexplained facets of their distribution emerge. We begin by considering variation in numbers of species and individuals, or diversity.

Benthic diversity

Diversity of the benthos varies greatly both with geographical region and with variation in physical conditions. One of the first attempts to explain such variation was made by Sanders (1968), who took dredge samples from muddy bottoms over a wide range of depths off the east coasts of North and South America, and around the coasts of India. He plotted the numbers of species found in each sample against the number of individuals, and interpolated what he called 'rarefaction curves' (Fig. 6.1). Some surprising points emerged from this wide-ranging study. Notably, both tropical shallow water and temperate deep water showed high values of diversity, while temperate shallow water and estuaries worldwide showed low diversity. It might seem unlikely that either the high diversity or low diversity areas have much in common. But Sanders suggested that the two groups are characterized, respectively, by long-term physical stability and by physical variability. Thus in tropical shallow water and the deep ocean, physical conditions such as temperature, salinity, and water movements are stable over long periods of time, allowing biological relationships such as competition, predation, and food-web interactions to determine community structure. Under these conditions, large numbers of species might evolve. In contrast, shallow-water areas and estuaries have physical conditions

Fig. 6.1 Rarefaction curves showing ranges of the calculated relationships between numbers of polychaete and bivalve species, and the number of individuals collected, in a variety of sublittoral sediments. (After Sanders, 1968.)

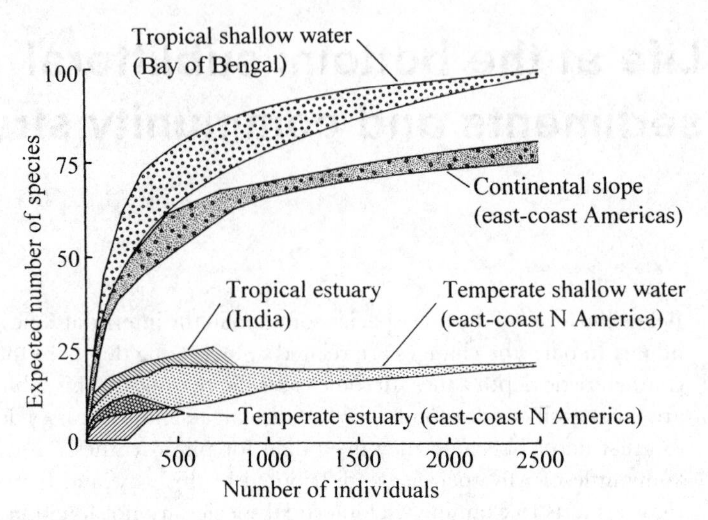

of salinity and temperature that fluctuate widely over time, and are critical in stressing communities so the development of complex biological interactions is prevented. Under these conditions, species numbers remain low. Because of his emphasis on contrasting degrees of stability over time, Sanders called this hypothesis the 'stability-time hypothesis'.

More recent studies have shown this to be too simple a picture. For example, if stability over time is the major determinant of high diversity, there should be more species on the abyssal plain (about 4000 m) than on the continental shelf. But for most faunal groups, maximum diversity is found not in very shallow waters, nor at great depths, but at intermediate depths of 1000–2000 m (Ormond *et al.*, 1997). Even this is not a rule, however—decapod crustaceans are most abundant in the top 500 m, while holothurians (sea cucumbers) are equally abundant at 1000 m and 4000 m. A variety of studies have also shown that diversity is very variable from place to place, with a continuous spectrum rather than polarization into high diversity and low diversity areas.

The stability-time hypothesis has, in particular, been criticized because now that Sanders' original data have been supplemented by later surveys, they are seen not to be entirely representative (Gray in Ormond *et al.*, 1997). Thus studies on shallow coastal faunas in soft sediments on the Norwegian shelf and around Australia have shown diversity to be much higher than Sanders found off east-coast America, and indeed higher than his values for the deep sea. It may be, then, that the overall idea of diversity increasing with depth needs reassessment. Nevertheless, there is also a counter-argument that suggests the original figures for the deep sea were too low! The arguments are evidently set to continue until more evidence accumulates.

In any case, it should be stressed that the distribution of many species cannot always be strictly related to depth because animals occur in patches of various sizes rather than in zones. The occurrence of these patches has fascinated marine biologists for decades, and one of the major aims of those working in the sublittoral is to be able to explain how patches come about. We spend most of this chapter discussing them, and how the animals that are found within them should be regarded.

Sublittoral communities: what governs species distributions?

In 1914, the Danish marine biologist Johannes Petersen began publishing the results of work in which he sampled the fauna from 193 sublittoral sites between Denmark and Sweden, at the mouth of the Baltic Sea. He realized that groups of species often occurred together, and using individual species to give these groups names he described seven major types of 'community'. For example, he described a *Macoma* community which was characterized by the bivalve *Macoma balthica*, but also contained other bivalves such as *Mya* and *Cerastoderma*. Petersen's community scheme was refined by Thorson (1957) who also worked in Danish waters, but expanded the scheme to many other parts of the world. The approach has been widely accepted, and many marine biologists now talk about benthic communities in the sense of those described by Petersen and Thorson.

The community approach is very similar to that describing 'associations' of plants on salt marshes, described in Chapter 5 (p. 117), in which, for instance, an area dominated by the cordgrass *Spartina* is called a Spartinetum, and similar problems are attached to it (J.S. Gray, 1981). Do such communities, with sharp boundaries between them and the next community, really exist? Are the communities to be regarded as mere statistical entities, in which species occur together because they are both governed by the same physical conditions? Or do the component species really interact, creating a tightly woven biological unit?

Persistence of community types and relation to substratum composition

Species and individuals in soft sediments show neither a random distribution nor a uniform one. Instead, they show patches of abundance that range enormously in size. At the bottom end of the scale, the distribution of protists may vary over distances measured in microns, while at the upper end, some benthic macrofauna may occur in patches that stretch for hundreds of kilometres. Often the observed patches do not seem to coincide *in detail* with changes in sediment characteristics, yet as we have emphasized in earlier chapters, many infaunal species are generally found only in particular types of sediment. Some burrowing amphipods, for instance *Haustorius* spp., are found in clean sand; while others, like *Corophium volutator*, occur only in mud. The same holds true for sublittoral fauna, so it is helpful here to return briefly to the ways in which size fractions of sediments are classified (see p.13). One way of visualizing sediment composition is to plot the various proportions of fine particles, sands, and gravels on one diagram (Fig. 6.2). From this it can be seen that while some sediments are relatively uniform in composition, others are very mixed. In general terms, each sediment type can be said to have its own typical fauna.

Off the north coast of France, for instance, sandy gravel usually supports the bivalve *Glycimeris*. Medium sand has the polychaete *Ophelia*. Muddy fine sand has both the bivalve *Abra alba* and the polychaete *Pectinaria*, while mud has *Macoma balthica* (Thiébaut *et al.*, 1997). But using these 'characteristic' species to name each community hides the fact that these substrata support over 100 other species as well, and both these and the characterizing species fluctuate over time. Thus, although the *Abra–Pectinaria* community as a whole persists in the same locality over many years, many of its component species show wide changes in distribution when examined even over 4 years. Over longer periods, quite drastic changes can be seen. In a 15-year study of four separate sites off the French coast, the 'characteristic' species *Abra alba* constituted as much as 15% of some communities but less than 1.5% of others (Fromentin *et al.*, 1997). Meanwhile the total number of species at each site varied from 130 to 420.

Observations made during monitoring of intertidal infauna suggest that here, too, communities can persist over long periods of time, even though they show some changes in species composition. In Manukau Harbour, an area of intertidal sand flats in New Zealand, communities of bivalves and polychaetes at six sites retained many of their dominant species over a period of more than

Fig. 6.2 Composition of shallow-water sediments from the Bay of Brest, northwest France, plotted as proportions of fine particles, sands, and gravels. Subjective grouping allows the distinction of five categories, but each has a wide spectrum of composition. Black circles show individual stations. (After Grall and Glémarec, 1997.)

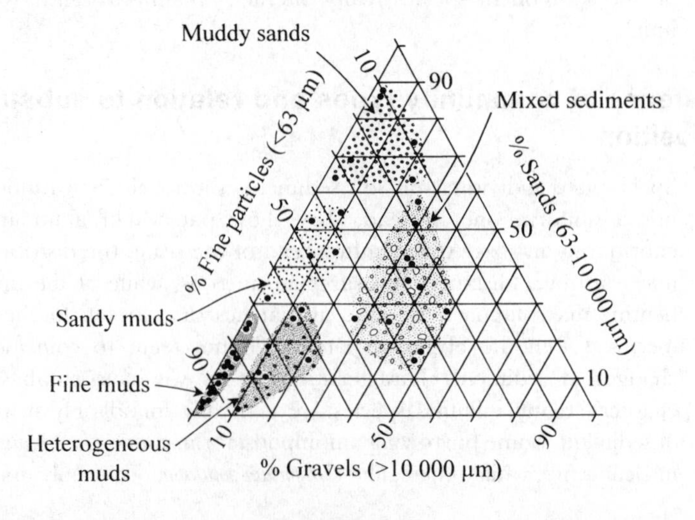

5 years, and were thus termed 'persistent' (Turner *et al.*, 1995). But they also showed significant fluctuations, some of which involved short-lived species whose recruitment and mortality led to sequential changes in species dominance. Disturbance from wind-generated waves, on the other hand, was estimated not to disrupt communities as would be predicted from the stability–time hypothesis, but to maintain them: turbulence and disturbance of the sediment helps to transport post-settlement stages of infauna, and thus encourages recolonization throughout the area.

If communities change in composition over time, a *close* relationship between community composition and substratum type is unlikely. And indeed we have already emphasized (p. 27) that there are many other reasons why this should be so. For instance, the sediment reflects factors such as benthic shear stress, and it may be that animals respond directly to the shear stress, so their link to the sediment would be an indirect one. Organisms themselves affect and alter sediment properties, so over time conditions may gradually become more appropriate for species other than the original colonists, but there is likely to be a time-lag effect here. Direct biological interactions between species may also affect community composition. One approach to these various problems is to consider that effects on community composition can be divided between recruitment limitation (i.e. larval supply and larval choice) and post-settlement processes (i.e. biological interactions and physical effects upon adults), as we have discussed on p. 30. Although there is little evidence concerning the relevance of variation in recruitment in explaining sublittoral distributions, it is felt that this, termed 'supply-side ecology' must in fact be important. We begin with this approach, and then focus on the various ways in which disturbance can affect adult patchiness.

Patchiness: the influence of recruitment

Pelagic larvae have a range of adaptations to a planktonic existence. At one end of the spectrum are 'planktotrophic' forms, which feed in the plankton and usually float there for a considerable time—weeks or months. At the other end are 'lecithotrophic' larvae which are provided with a yolk supply and do not have to feed while afloat. These spend only a short time in the plankton. According to Thorson (1950), adult populations that have planktotrophic larvae fluctuate more in biomass than those with lecithotrophic larvae. Thorson concluded this is because life in the plankton is particularly hazardous, so that a long planktonic life would lead to large variations in settlement or 'recruitment'. This hypothesis remains unproven for lack of evidence: measuring recruitment, especially on soft substrata, is exceedingly difficult. The few studies that have attempted to measure it have not provided evidence to support Thorson's views (Olafsson *et al.*, 1994). There is, however, evidence that the ways in which pelagic larvae are distributed may strongly affect adult distribution. In the *Abra* communities off northern France discussed in the last section, larvae tend to be retained within a close area because of weak residual currents.

But further north, in the North Sea, larvae are transported parallel to the coast as planktonic patches 10 km long, and these lead to spatial heterogeneity in recruitment (Thiébaut *et al.*, 1997).

Larval recruitment may also be affected by processes determined after the pelagic phase. When larvae settle on the bottom, competition between larvae of different species may result in differential mortality. The settlement of a polychaete, *Pectinaria*, for instance, is adversely affected by the presence of larvae of another polychaete, *Owenia* (Thiébaut *et al.*, 1997). This effect may be strengthened by adult–larval interactions. For example, adult *Owenia* increase mortality of settling *Pectinaria*, and many suspension feeders have been shown to inhibit larval settlement (Olafsson *et al.*, 1994). Experimental studies in the Rance estuary, northern France, have also shown that not only can predation on newly settled polychaete recruits reduce their number and biomass, but the presence of predators can make recruits move away, altering distribution patterns (Desroy *et al.*, 1998).

Overall, the influence of 'supply-side' ecology on soft-sediment communities has been much neglected, and it may be that future work will bring to light more examples in which patchiness is related to larval dispersal.

Patchiness: the influence of biological disturbance

Currently, many biologists think that one of the underlying causes of patchiness is disturbance—either direct physical disturbance, or more often because of effects caused by other organisms. In Chapter 2 we briefly discussed the ways in which animals disturb the sediment—burrowers loosen it, predators create pits in the surface, and so on. These disturbances can occur on a range of scales, and consequently it takes various times for the habitat to recover (Fig. 6.3). Most benthic infauna do not disturb more than a few square millimetres, and such disturbances disappear within a day, whereas gray whales, winnowing the sediments for amphipods, may disturb areas of several square metres, and these do not recover for months. Some of the larger influences, including anoxia and the effects of fishing, may last for years. We consider in this section some examples of the ways in which organisms can cause localized effects, and go on, in the next, to look at physical disturbances.

Sublittoral sediments often present a rather uniform appearance, in which the only obvious features are crustacean burrows. Do these, or their occupants, influence the distribution of the infauna? In the shallow sublittoral sandbanks of Otago Harbour, New Zealand, the only common burrowing crab is *Macrophthalmus*. When the area around *Macrophthalmus* burrows is disturbed experimentally, recolonization over time can be examined (Fig. 6.4). Deposit feeding polychaetes soon recolonize areas 1 m away from the burrows, but few individuals of any species are found right next to the burrows, even after a month (Thrush, 1988). The burrows thus form the centre of low-density patches of infauna. In this case it is probable (but not proven) that the physical disturbance

Fig. 6.3 The sizes of disturbance patches created by a variety of mechanisms, and their relationship to recovery time. Stipple shows the spread of values caused by anoxia. (After Hall *et al.*, 1994.)

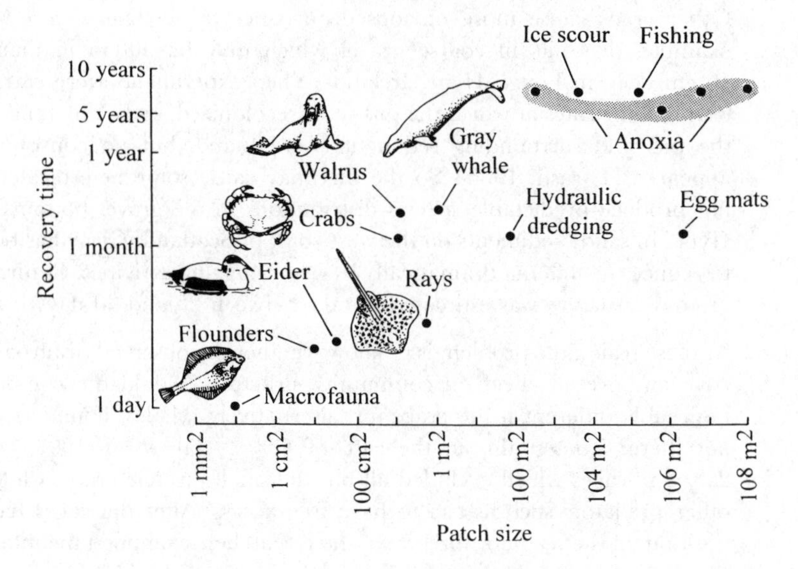

Fig. 6.4 Recolonization of sublittoral sands by two polychaete species in relation to burrows of the New Zealand crab *Macrophthalmus hirtipes*, after mechanical disturbance. Clear histograms show densities near burrows, hatched histograms show densities away from them. Error bars show 90% non-parametric confidence intervals. (After Thrush, 1988.)

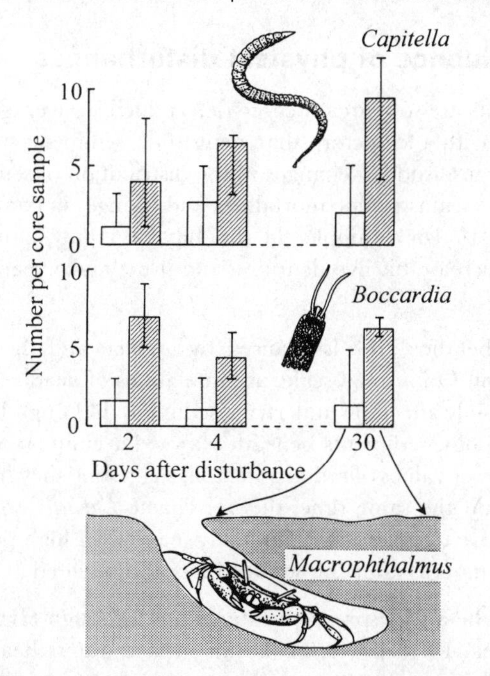

of the crabs, walking in and out of their burrows, causes the effect—rather like human trampling on pathways over the beach.

Other crabs cause more obvious disturbance by feeding. *Cancer pagurus*, for example, digs pits in coarse gravel which may be 30 cm in diameter and 10 cm deep in Lough Hyne, Ireland. When experiments were carried out to examine the rates at which the pits were recolonized, some differences between the pits and surrounding sediments were found, but no consistent pattern appeared (Thrush, 1986). So the pits may cause some heterogeneity, but do not produce predictable species distributions in the gravel bottoms of Lough Hyne. In sandy sediments on the west coast of Scotland, *Cancer* has been shown to reduce the infauna dramatically by its pit-digging activities. Return to 'background' structure was estimated to take between 25 and 30 days.

At these scales, the problem is to know whether the observed disturbances really have any overall effect on community structure. Would the communities be noticeably different if the crabs (or, alternatively, whales, flounders, etc.) were not there? To test this in the case of *Cancer*, Hall *et al.* (1993) constructed 2 × 2 m cages which excluded all but the small juvenile crabs, while allowing other predators such as fish to have free access. After the cages had been in position on the sea floor for 1 year, the researchers examined the infauna inside them and compared it with randomly selected areas outside. They could find no evidence for any difference between the two. In this case, it seems likely that despite the immediate effects of crabs on small areas when they dig pits, there is no large-scale effect upon the community as a whole: other, possibly larger-scale processes, must be determining community structure.

Patchiness: the influence of physical disturbances

Bottom sediments are to a great degree determined by average current speeds, in conjunction with the factors that determine sediment supply. But storm conditions can cause sudden changes in the distribution of sediment, and these may in turn cause catastrophic mortalities and changes in community structure (Hall *et al.*, 1994). For example, in the Bay of Fundy, storms remove fine sediment and decrease bivalve densities, and it may take years before communities recover.

Effects on the benthos are also caused by anoxia, and these are extremely variable in extent. On a small scale, accumulations of dead seaweed only 1 m^2 in extent can rapidly affect the underlying sediment. In Lough Hyne, a small and enclosed water body, sediments beneath seaweed accumulations become more acid and have lower values of redox potential, suggesting they have lower oxygen concentrations. At the same time, the polychaete *Capitella capitata* increases in abundance relative to other areas, and may maintain a high population density for as long as 2 months after the seaweed has accumulated (Thrush, 1986).

Anoxia can also be a widespread phenomenon. In Lough Hyne, a thermocline forms in summer at a depth of 20–30 m: surface waters heat up and as they become less dense, mixing with lower layers ceases. The lower layers then progressively use up available oxygen, until the water becomes virtually anoxic, and rich in toxic sulphides (Kitching *et al.*, 1976). As a result, most macrofauna below the thermocline die. A few tolerant bivalves such as *Corbula gibba*, survive, but even the very tolerant polychaete *Pseudopolydora* is eliminated by late summer. Recolonization of the lower layers occurs in autumn when the thermocline breaks up and bottom waters once again become oxygenated. The dominant species are opportunists—species that disperse, settle, and grow rapidly. *Corbula* and *Pseudopolydora* are abundant, but there is also *Abra* (another bivalve) and the polychaete *Capitella*. The composition of communities below 25 m is therefore largely determined by the summer anoxia. It is also possible that the anoxic layer below the thermocline may wash over slightly shallower areas of the lough at times, causing temporary kills of macrofauna. It is not surprising, therefore, that the communities at this depth are all dominated by species with 'low successional status'—there is never time for successions to develop very far before they are re-set to zero by anoxia.

On a wider scale still, anoxia caused by organic enrichment or 'eutrophication' is now common in the marine environment (Diaz and Rosenberg, 1995). Wide-scale anoxia is largely generated indirectly through human agency: increased nutrient input (e.g. from sewage or land run-off) stimulates increased primary production, and the dead remains of the plants that result then use up oxygen as they decay. Anoxia has now been recorded from such widely dispersed areas as, for example, the Chesapeake Bay (USA), the Baie de Somme (France), and the Baltic Sea. In the Baltic, where an enclosed basin is connected to the ocean only by very narrow sounds, and where the drainage catchment area supports more than 70 million people, rapid changes have been seen since the 1960s (Bonsdorff *et al.*, 1997).

How do varying degrees of eutrophication affect benthic community structure? Studies of a large number of situations suggest a complex interplay between species numbers, total biomass, and total abundance of individuals (Fig. 6.5). In normal (unpolluted) areas, there are usually many species, but total biomass and density of individuals are relatively low. Here the re-working of the sediment, including burrowing by crustaceans such as *Nephrops*, keeps the sediment aerobic to a considerable depth. With increasing organic load, species numbers, biomass, and abundance all rise, but the depth to which the sediment remains aerobic decreases. In this 'transitory' zone, bivalves such as *Corbula* and polychaetes such as *Pectinaria* (*Lagis*) may be common. In more strongly polluted areas, abundance rises but species numbers and biomass fall; and in grossly polluted areas there are no macrofauna, although there may well be meiofauna typical of the sulphide system (p. 17).

Abundance in polluted areas is thus governed by the influx of only a few species. Are these the most tolerant of anoxia? The answer is that while species in organically polluted zones are indeed tolerant of anoxia, they are *not* the most tolerant species. Thus the most universal 'indicator' of high organic pollution is

Fig. 6.5 Changes along a gradient of organic enrichment. The top diagram shows some of the infaunal species and their relation to aerobic and anaerobic sediments. The lower diagram shows a generalized picture of species numbers, total biomass and abundance of total individuals. (After Pearson and Rosenberg, 1978.)

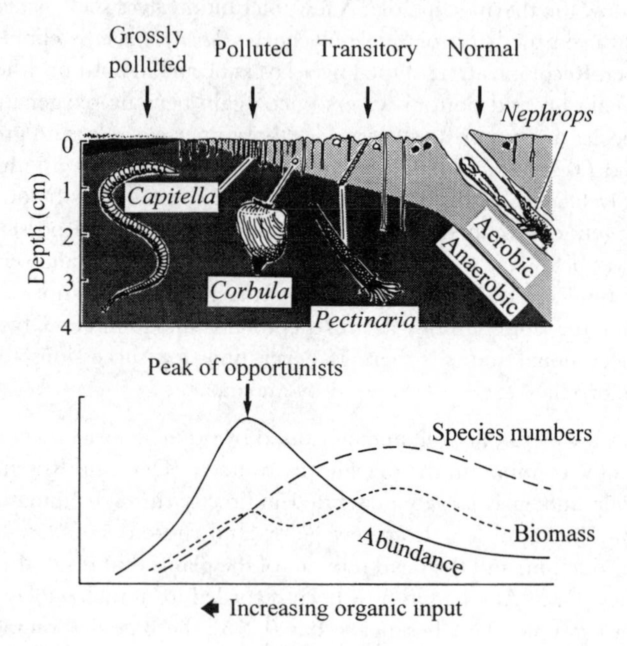

the polychaete *Capitella capitata*. But several other species, such as *Arenicola* and *Corbula*, are in fact more tolerant of severe anoxia (Diaz and Rosenberg, 1995). The explanation may lie in the fact that the colonizers of polluted environments are 'opportunists'—those species that can reproduce and grow rapidly, taking immediate advantage of any sudden environmental change, where competitors are removed (Pearson and Rosenberg, 1978). The observations in polluted gradients therefore suggest the idea of a succession of species rather than a series based entirely on tolerance: anoxia destroys complex communities, opportunists move in, and then, given time and the subsequent reduction in the source of anoxia, complex communities can develop again.

The idea of successions is reminiscent of discussions concerning community structure on salt marshes (p. 107). But as we concluded for marshes, successions by themselves do not explain present-day situations. We need to look at the ways in which successions may vary both in time and space, and the scales on which they may be re-set by disturbance.

Patchiness: can disturbance explain distribution patterns?

The result of disturbance on any scale is to alter conditions on the bottom—particularly for the infauna. In some cases the entire community will die. In others, some species will be reduced in abundance, leaving more space or resources for the remainder and for incomers. Recovery then proceeds partly by settlement of larvae, partly by influx of adults and juveniles: recolonization by adult amphipods, for example, frequently occurs. Either way, it will take some time for a patch to recover its former characteristics of species composition and abundance, because the colonizing community usually exhibits some form of succession, as we have just discussed: opportunists arrive first, grow, and reproduce rapidly, and only later will the species that mark a more stable community arrive.

Over a wide area, then, one might expect to find many patches at different stages of succession, reflecting differing times at which they were disturbed. This model of 'patch dynamics' is now widely accepted as an explanation, not just for the patchiness of benthic soft-bottom communities, but for equivalent community structures on rocky shores, in terrestrial forests, and other habitats.

How good is the evidence for this explanation? Unfortunately, the experiments and observations on patch formation and recolonization have so far been on quite small scales—yet some of the observed patchiness is on an enormous scale. We have already discussed the apparent lack of effect on the community of pits dug by crabs. But disturbances may occur on many scales, as emphasized earlier in the chapter. To take just one example, Lough Hyne: crabs dig pits less than 1 m^2; seaweed accumulates over areas of 1–50 m^2; and the thermocline may spill over areas in excess of 1000 m^2 (Thrush and Townsend, 1986). Which of these, if any, are responsible for the observed patchiness?

The theory of disturbance as a prime factor in causing mosaics or patchiness in soft sediments derives in great part from studies on rocky shores (Little and Kitching, 1996). But rocky shores and soft sediments are very different environments as far as organisms are concerned (p. 2). Future studies on sublittoral sediments will have to demonstrate that the scale of disturbances matches the scale of the mosaics if disturbance theory is to be validated. It also needs to be integrated with observations of variation in larval settlement, as these may also produce patchiness, and the study of supply-side ecology is still in its infancy.

Techniques

Working in the sublittoral poses a whole series of extra problems for any biologist used to working on the shore between the tidemarks. Taking samples ceases to be a matter of using spades and quadrats, carrying out experiments becomes a technical nightmare, and even knowing where you are is no longer easy. We briefly discuss some of the ways round these problems. Assuming one has been able to gather some information, there is then the problem of how to analyse the data.

Sampling methods

Most sublittoral sampling is undertaken from boats, using equipment such as grabs or corers (Holme and McIntyre, 1984; Hartley and Dicks in Baker and

Wolff, 1987). Grabs usually operate with cylindrical jaws that bite into the sediment, so while they sample a known area, they have the disadvantage that this area is sampled to varying depths. As they seldom bite deeper than 20 cm, they will miss the deeper-burrowing infauna. The simplest grabs, such as the Van Veen, are most commonly available, but often misfire or topple sideways, thus wasting valuable sampling time. More complex, expensive, and heavier varieties like the Day grab are more reliable but cannot be used from very small boats. Choice of grab is therefore often dictated by a combination of funding and size of available boats. This is unfortunate, as different grabs have different sampling efficiencies. While these are hard to evaluate, they are crucial!

Corers are either cylindrical or square tubes that are thrust vertically into the sediment. They are particularly useful in mud, because they can be brought to the surface without the need to close the bottom end. In loose sand, though, more sophisticated devices are necessary because the sediment tends to fall out of the coring tube. When corers are used, they penetrate more deeply than grabs. They are often appropriate for meiofauna sampling, with small diameters (of the order of 20 mm).

Neither grabs nor corers will sample mobile epifauna properly. A wide variety of semi-quantitative trawls, 'epibenthic sledges', and other devices designed to fish at the sediment-water interface have been constructed, but are specialist items.

In addition to the technical problems of taking a quantitative sample is the problem of accurately plotting its position. With the advent of satellite-based global position-fixing systems, this is becoming easier (Riddy and Masson in Summerhayes and Thorpe, 1996), but is expensive. Modern systems can pinpoint positions as accurately as 1 m.

When working sublittorally, most sampling is done remotely, with all the inconvenience of not seeing what is happening at the sample site. One way of getting round this problem is to use SCUBA equipment. Free diving is limited to about 40 m, but within this limit it allows much greater flexibility—meanwhile imposing its own problems! Samples can be taken from accurately determined positions, and experiments can be set up and monitored. But diving is very time-consuming and constrained by bad weather (Hiscock in Baker and Wolff, 1987), and for long-term studies is best carried out in sheltered areas.

Overall, sublittoral researchers have to be satisfied with less data for a far greater expenditure than those working on the shore. On the other hand, they have enormous areas to investigate: virtually all areas of the ocean below the level of low tide are covered by soft sediments.

Analytical methods

This book does not cover statistical methods in detail. Nevertheless, it is appropriate to mention in general terms one of the major problems in dealing with the fauna of sublittoral sediments, and that is the large numbers of species and individuals that are present. How does one analyse variations in patterns of distribution which involve thousands of individuals and hundreds of species in each square metre? In particular, we have already discussed the question of whether sample stations are clustered into 'communities' or whether they show infinite gradations of species composition.

Two major types of approach have been used (J.S. Gray, 1981; Baker *et al.* in Baker and Wolff, 1987). One common technique is to use the dendrogram, in which similarity between samples is calculated and then plotted as a branching diagram. Here the ends of the branches—or any length along them—can then be compared in terms of percentage similarity. The second common approach is called ordination, in which stations are first plotted as positions in multi-dimensional space. The multi-dimensional plot is then represented in two dimensions. Stations that are closest together in this plot have greatest similarity. Both approaches allow some estimation of the degree to which fauna can be placed in communities, or should be regarded as continua.

7 Estuarine habitats and coastal lagoons

Sediments from coarse gravel to fine mud can be found in varying degrees both on relatively open coasts and in shelter. The greatest reservoir of sediments is in the sublittoral zone, stretching down to the deep sea. But the most easily accessible and most varied accumulations occur in estuaries, so not surprisingly a great majority of soft-sediment researchers have concentrated their efforts there. Estuaries are more than just indentations in the coastline, however, and provide a variety of characteristics that differentiate them from strictly marine environments.

Part of the fascination of estuaries lies in their dynamic nature—most physical variables really *are* variable! Take the most commonly measured variable, salinity, for example. Over an entire estuary this will vary from fresh water to full-strength sea water or above, and may fluctuate considerably even at one site with different stages of the tide and with different seasons. Besides this, differences in tidal regimes may produce strong currents and turbid water. In the Bay of Fundy, for instance, where maximum tidal range can exceed 16 m, currents average 0.7 m/s, and concentrations of suspended sediment are usually in the range of 0.1–1 g/l (Gordon, 1994). In the Severn estuary, average currents may exceed 2 m/s, and extremes reach over 4 m/s. Sediment concentrations near the bottom may achieve quite extraordinary levels, exceeding 200 g/l and forming what is known as 'fluid mud'. The fine sediment load of the Severn at spring tides exceeds 30 million tonnes (Kirby, 1994). Many estuaries also show much higher levels of organic material than do open coasts, and they often have a great variety of habitat types. This diversity adds another dimension to the interest of estuarine studies. In the Severn estuary, for example, there are numerous places where salt marsh, mud, sand, gravel, and rock can all be found within 1 km^2. Because of the tremendous physical variations within estuaries, they produce a rigorous environment for fauna and flora, which have to adapt structurally, physiologically, or behaviourally in order to survive. As we shall see, there are relatively few organisms that can survive within an estuarine regime, but those that do can often flourish.

What are estuaries?

Estuaries are thus places of much variation in both time and space, raising the question of how to define them—in other words, what are the crucial points in all these variable factors? The word 'estuary' is derived from the Latin *aestus*, meaning tide, so this must figure in any definition. Dictionaries, however, centre on the meeting between the sea and a river, and physical oceanographers usually take this to imply that there is mixing of sea water and fresh water, with consequent production of intermediate salinities. In arid regions, however, bodies that seem otherwise to fit the term 'estuary' may have salinities *higher* than that of the sea. Here, therefore, we define an estuary as 'a partly enclosed tidal inlet of the sea in which sea water and river water mix to some degree'.

Any definition will be hard to defend at the extremes of its range. In the case of estuaries, there is a wide spectrum of types, particularly in terms of the degree of connection with the sea. At the 'open' end of this spectrum, estuaries are indistinguishable from bays of the sea into which small rivers run, because these would produce hardly any change from a strictly marine condition. At the 'closed' end of the spectrum, estuaries merge with coastal lagoons, in which mobile sand and shingle deposits isolate lagoon water from the sea. Indeed, as some lagoons have only seasonal openings to the sea, they may be said to change status from estuary to lagoon over the year! In addition, rivers that run into tideless seas like the Baltic produce environments very like classic tidal estuaries, so it is not surprising that many authors have concluded it is unlikely that any one definition of estuaries will satisfy everyone.

Instead of trying any further to confine estuaries within a form of words, we therefore go on to the more profitable exercise of considering some of the characteristics of estuaries and their variability, and how this variability provides a range of habitats for fauna and flora.

Estuarine tides

Almost all estuaries are tidal areas, but the degree to which tides dominate over other processes varies widely. As we shall see later, this variation in tidal forces is extremely important in determining estuarine characteristics.

In Chapter 1, it was pointed out that the rise and fall of tides in estuaries can be very different from the shallow sinusoidal curve found in oceanic situations. How do estuarine tides affect conditions as far as the biota are concerned? Two properties of the tide-wave are important here—one is the amplitude of the wave, and the second is its asymmetry.

Tidal amplitude

Tidal range along any coastline can be extremely varied, due to the way in which ocean tide waves are modified by the proximity of land masses. In enclosed seas like the North Sea, coastal tides can vary in range from 2 m to 6 m within distances as short as 100 km. On top of this initial variation, the tidal range within estuaries is extremely variable. One might expect tidal range to decrease inside estuaries, because of the increased friction between water and the estuary's boundaries. In a few estuaries this effect can be seen, but in most it is offset by the effects of the convergence of the estuary's shores. This convergence concentrates the energy of the tides and thus increases their amplitude. In the Bay of Fundy, for instance, the range is 5 m at the mouth but this rises to 9 m before the bay splits into two basins. Here the range increases to 15 m and occasionally to 16 m. In the Severn estuary, a range of 7 m near the mouth builds to about 14 m upstream before diminishing again (Fig. 7.1). While these values are extreme, the picture of increased tidal range in the middle reaches of estuaries is probably the most widespread one.

As estuarine tide waves move through relatively shallow water, their movement involves not just an up-and-down one as discussed for the open sea (p. 9), but a lateral displacement of water. Just above Avonmouth in the Severn estuary (Fig. 7.1), nearly 1 km^3 (actually 700 million m^3) of water moves upstream and then downstream on a spring tide. Not surprisingly, therefore, current speeds are high and the consequent scour makes conditions difficult for the biota. Not only are bottom sediments unstable, but suspended sediment loads are high and vary tremendously over the tidal cycle. In contrast, conditions in estuaries that have a reduced connection with the sea may be relatively placid. Some coastal

Fig. 7.1 Spring tides in the Severn estuary. A symmetrical tide curve at the mouth (Milford Haven) is amplified at Avonmouth, then declines and becomes progressively more asymmetrical upstream (Sharpness), as the lower part of the wave is cut off by the rising bed of the estuary. At Minsterworth the front of the tide wave forms a bore. (Partly after Rowbotham, 1964.)

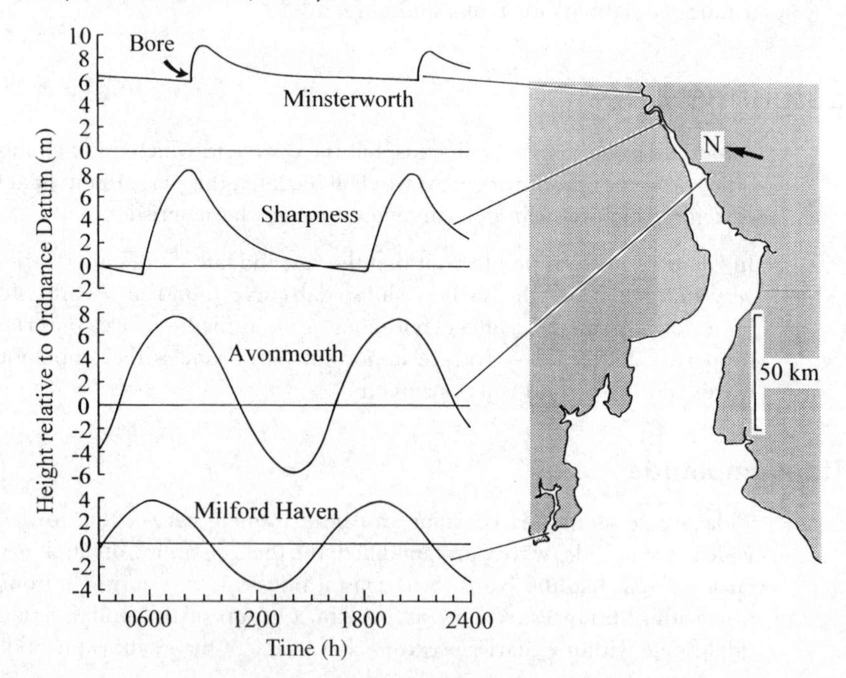

lagoons have either no tide at all, or a very slow rise and fall due to seepage through a shingle bank.

Tidal asymmetry

As a tide wave moves up an estuary, friction at the bed reduces the velocity of the lower part of the wave, and consequently the shape of the wave changes. Instead of being a symmetrical sine wave, the time of flood (upstream) is shortened while the time of ebb (downstream) lengthens, and a kind of asymmetrical saw-tooth pattern is produced (Fig. 7.1). One of the major consequences is that the high-velocity flood tides bring in sediment which the lower-velocity ebb tides do not remove. Estuaries therefore often act as sediment traps, gathering deposits from the sea as well as from the rivers that feed them. The distribution of much of this sediment is also related to the movement of the tides: flood tides tend to carve their own channels facing upstream, while ebb tides form river-like channels facing downstream. The deposited material may even be re-aligned between each ebb and flood, making an extremely unstable environment for the infauna.

In extreme cases, where the tidal range is very high, tides can become so asymmetrical that the flood tide moves up-estuary as a 'wall' of water called a tidal bore (Fig. 7.1). Bore waves can be extremely destructive, causing severe bank erosion, and thus reducing the options for benthic infauna. In the Severn, for instance, the bore can be over 1 m high and can travel at 5.8 m/s. In the Amazon, the bore can be 5 m high.

Types of estuary

Because estuaries are so variable, it is convenient to impose some kind of classification upon their great range of forms, despite the inevitable problem that they form sequences that merge into each other. But classification schemes depend to a great extent upon the bias of the classifier: geographers may be mainly interested in origins, hydrographers concentrate on water movements and composition; while sedimentologists are more interested in depositional processes and the factors that control them. We take three classifications in turn, and discuss which are most useful for biologists.

Origins: how were estuaries formed?

The shape of an estuary on the map is usually a good guide to the way it came into being (Fig. 7.2). Thus 'drowned river valleys' show the branching structure of old river systems. These, the most common kind of estuary, were formed after the last ice age as sea level rose to its present level about 6000 years ago (Pethick, 1984). They are sometimes called 'coastal plain estuaries' and often contain large areas of mudflat and salt marsh. Examples include many on the east coast of the USA, such as the Chesapeake Bay, and in southwest England, such as the Severn. Both these two examples are large enough

Fig. 7.2 Three types of estuarine body as seen on the map. Drowned river valleys (e.g. the Chesapeake Bay) have intricate dendritic structure. Note also the offshore barrier island on the coast to the east. On this diagram is shown the change in salinity along the bay. The isohalines (lines of equal salinity) are further to the right, facing downstream, because of Coriolis effects. The Chesapeake Bay is large enough to have many sub-estuaries entering it. (After Lippson and Lippson, 1997.) Fjords have much simpler, parallel sides (e.g. Ronas Voe, Shetland). Barrier islands may form lagoons between themselves and the mainland; the Huizache and Caimanero lagoons of the Pacific coast of Mexico also have complex links with rivers, which retain connections with the sea. (After Edwards, 1978.)

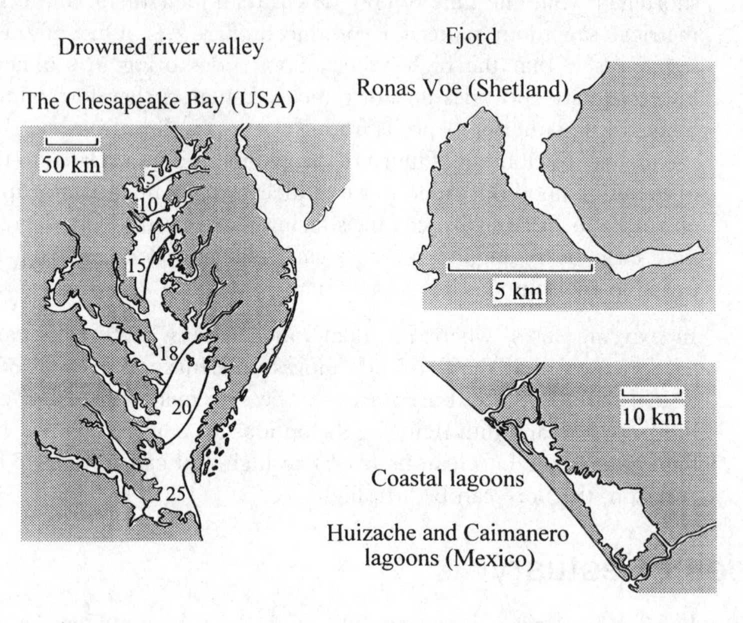

for the major drowned valleys to have tributaries, and these are often referred to as sub-estuaries.

The second group in this classification are the fjords, which were also formed by rising sea level, but began as glaciated valleys rather than river valleys. They are usually deep with very steep rocky sides, and often much straightened in comparison with drowned rivers—good examples are found in Norway and on the Pacific coast of Canada. Fjords contain much coarse sediment brought down by the glaciers, and often this material has been deposited in the form of 'moraines'.

The third common type is the bar-built or barrier-built estuary. These occur at river mouths where sand and shingle has accumulated offshore to form islands or spits. They may form long systems of interlinked estuaries or lagoons, as on the Gulf coast of Texas and the Wadden Sea on the coast of the Netherlands and Germany. Such estuaries are shallow and it is here that the cut-off points between estuaries, coastal bays, and lagoons are most difficult to define.

How useful is this classification for biologists? Some features of each of the three estuarine types are of direct concern to the fauna and flora. For example, drowned river valleys usually contain large areas of 'wetland'—regions where the soil is permanently saturated, and ranging from salt marshes to freshwater marshes and swamp forests. In the Chesapeake Bay, for example, there are more than 4000 km^2 of wetlands, including bald cypress swamps formed by *Taxodium distichum*. Drowned river valleys are therefore often places rich in interest to wetland conservationists. Fjord systems form a stark contrast because they usually have very little intertidal zone, or surrounding wetland. If they have sills, much of the biological interest is limited to surface waters. Bar-built estuaries form a very wide category, and perhaps provide most biological interest at their extremes—in fact, where they become coastal lagoons! Here, fringing macroflora may be extremely abundant. In the tropics, mangroves will dominate and in temperate zones salt marsh may fringe lagoons. In addition, there are often submerged macrophytes such as the pondweeds, *Potamogeton* and *Ruppia*. Because lagoonal substrata are entirely composed of mud and sand, burrowing animals form the dominant benthos; but fish, prawns, and other mobile species use this benthos as a food source. While these species may be of limited diversity, they may reach extreme levels of abundance, often exploited in the tropics for local fisheries. Some tropical lagoons are managed for the culture of fish, and in this case their inflow and outflow may be highly modified.

Mixing and movement of water

The most widely used classification for estuaries splits them up according to how much the sea water and river water mix. While mixing occurs to infinitely varying degrees, hydrographers usually recognize three separate classes, with a fourth found in the tropics. Of these, the 'salt-wedge' or 'highly stratified' estuaries show least mixing. They are found where a large river flows into an area with a small tidal range—a classic example being the Mississippi flowing into the Gulf of Mexico. Here the less dense fresh water floats on the denser sea water. As it moves seawards it spreads out, forming a thinner layer, while the sea water beneath assumes the shape of a wedge (Fig. 7.3). This wedge may move up and down the estuary slightly as the tide rises and falls, but its structure stays intact because of the minimal tidal mixing. On the other hand, it may move hundreds of kilometres as river flow increases or decreases. Some mixing of the sea water into the overlying fresh water does occur, because internal waves develop where the two water masses meet, and as these break upwards, salt water is 'entrained' into the fresh water. The process of entrainment reduces the volume of water in the salt wedge, and more sea water flows inwards to take its place, creating a 'residual' (i.e. non-tidal) current. The water in the salt wedge is thus gradually replaced so that it seldom runs out of oxygen.

The dominant river flow and small residual currents determine what happens to the river's sediment load. Most fine material is carried out to sea as a

Fig. 7.3 Distribution of salinity and sediments in salt wedge and partially mixed conditions. The top diagram shows the Mekong River estuary (Vietnam) during the wet season, where dominant river flow forms a salt wedge with a small turbidity maximum. The bottom diagram shows the situation in the dry season, where wind and waves mix the reduced river flow with sea water. Sediment previously deposited outside the estuary is returned and contributes to a large turbidity maximum. White arrows show water flow. Black arrows show sediment movement. (After Wolanski *et al.*, 1996.)

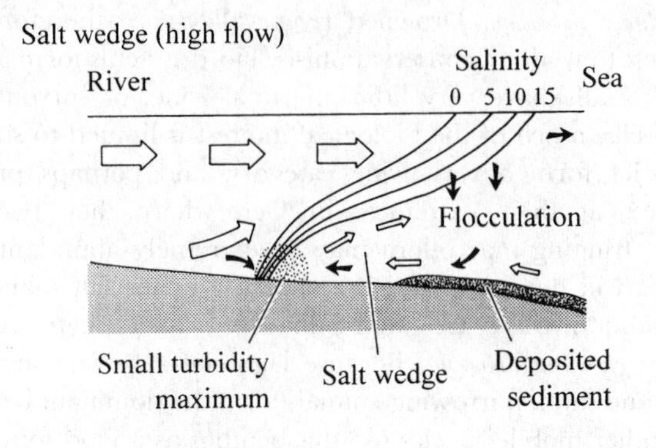

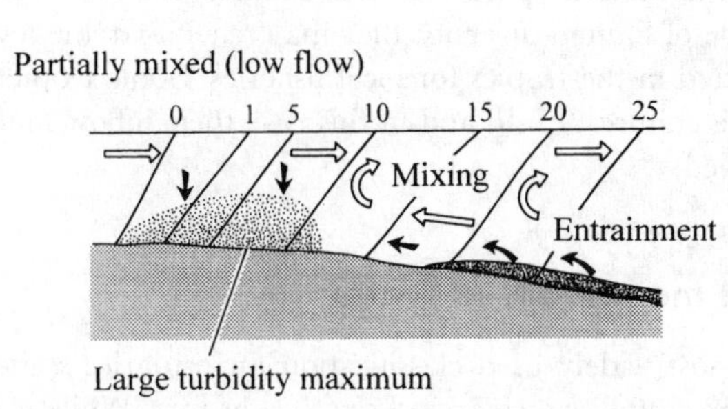

well-defined plume, and the estuary itself is dominated by coarse sediments. The coarsest, or bedload sediment, is deposited at the tip of the salt wedge as the river flow rises over the wedge and leaves the bed. This sediment may form a bar in the river—in the Mississippi, so much sediment may be deposited that the bar may rise by 2 m in a week. Fine sediment may also accumulate here, because when fine material precipitates through the wedge from the fresh water above, it is taken upstream by the residual currents. In the Fraser River estuary, western Canada, the highest turbidities actually form just *upstream* of the salt wedge (Kostaschuk *et al.*, 1992). Further downstream, the river's coarse suspended sediments are often deposited as a considerable delta.

Salt-wedge estuaries therefore form environments for organisms that show distinct transitions between regions: a vertical plankton tow would show a sharp transition from marine plankton in the wedge to freshwater plankton above it. Horizontal sampling of the benthos would show rapid transitions between sand fauna and the barren regions of coarse shingle, as well as transitions from marine species to freshwater species.

In terms of mixing, fjords fall into the same category as salt-wedge estuaries—they are usually highly stratified (Fig. 7.4). However, sills near the mouth prevent much circulation within the deeper water, so this may run out of oxygen. In consequence, the bottom waters of fjords are often anoxic and support little macroscopic life, although the sediments may be rich in chemosynthetic bacteria.

The second type of estuary in this scheme occurs where there is more tidal action, producing more turbulence, and river flow is smaller. These are 'partially mixed' or 'moderately stratified' estuaries, and here the mixing of river water downwards into sea water occurs in addition to the entrainment of salt upwards. The result is a gradual increase of salinity from surface to bottom, rather than an abrupt change as in the salt-wedge (Fig. 7.3). In this type of

Fig. 7.4 Stratification in a fjord and a negative estuary. The top two diagrams show Loch Ailort, a fjord in western Scotland. Salinity is plotted as isohalines (lines of equal salinity) which show the sharp boundary between dilute surface water and saline bottom water, caused by the sills. Oxygen falls to very low levels at depth because of this lack of mixing. (After Gillibrand *et al.*, 1996.) The bottom diagram shows the Spencer Gulf, a negative estuary in southern Australia. Here intense evaporation raises salinity at the head of the gulf. As this saline water cools in winter, it sinks and flows offshore, to be replaced by less saline surface water (arrows). (After Lennon *et al.*, 1987.)

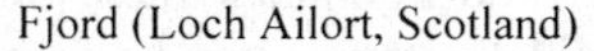

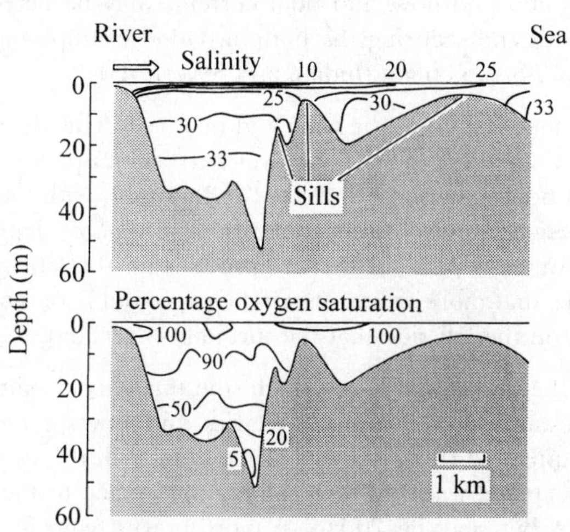

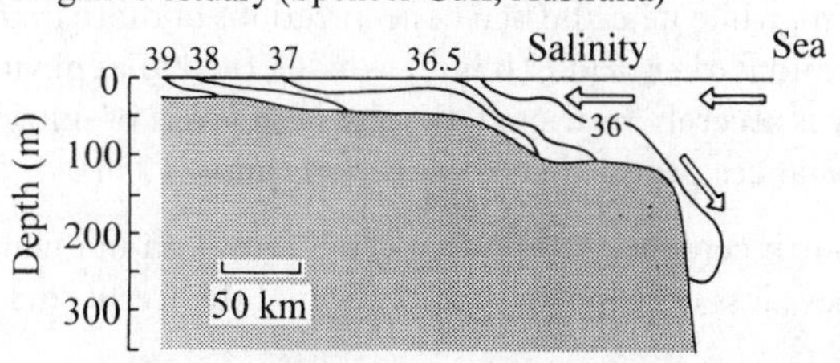

estuary, fine sediment does not usually flow out to sea as a plume. Instead, it sinks to the bottom, aided by flocculation (p. 59), and is then returned upstream by the residual currents. Fine sediment therefore accumulates within the estuary, and is most concentrated where upstream transport is balanced by the river's seaward transport, where it forms what is called the 'turbidity maximum'. This may be some kilometres in length, and will both move up and down the estuary and change in size as tides change between springs and neaps. In the Thames, for example, the maximum may move up and down by as much as 20 km.

The biological importance of the turbidity maximum is extreme. Fine sediment clogs the filtering mechanisms of suspension feeders, but maybe more importantly it cuts down the penetration of light through the water. This means that algal growth is suppressed, and as planktonic food chains mostly start with primary production, they may be much reduced in turbid estuaries. In the Ems estuary, the Netherlands, computer modelling suggests that a 50% reduction in turbidity would result in large increases of planktonic diatoms. In turn, substantial increases of copepods and other zooplankton are predicted (DeGroodt and Jonge in McLusky *et al.*, 1990).

In the third category of water mixing fall the estuaries known as 'fully mixed' or 'non-stratified'. Here the tidal currents are sufficient to remove all vertical salinity differences, and indeed tides dominate these estuaries. Currents reverse as tides ebb and flow, and tidal currents may be fierce, producing very high turbidities and scouring the bottom. Good examples are the Delaware River (eastern USA), Ganges (India), and Severn (UK).

Where these estuaries are wide and relatively shallow, the outward river flow is deflected to the right (in the Northern Hemisphere) by the Coriolis effect. Inward tidal flow is also deflected to the right, with the result that while there is no *vertical* salinity stratification, there is a *lateral* change in salinity (Fig. 7.1). Saline water is present further upstream on the left bank, and there is some evidence that more sediment may be deposited here than on the right side. It is usually on the left side that the turbidity is greatest.

When the tide range is very high, the turbidity maximum often spreads over great distances in fully mixed estuaries, and moves horizontally. This movement is accompanied by movement of the salinity distributions. For example, in the Severn, turbid water of high salinity may reach to the head of the estuary in summer, but may be 70 km or more downstream in winter. At spring tides, surface water with more than 0.5 g sediment per litre may spread over distances of 60 km, while near the bed concentrations of more than 3 g/l spread just as far (Kirby and Parker, 1983). It is no wonder, then, that phytoplankton growth in this estuary is severely restricted, despite high levels of nutrients that would in other circumstances produce eutrophication (Joint, 1984).

The fourth category of estuary occurs mostly in hot climates. In these 'negative estuaries', loss of water by evaporation at the freshwater end exceeds the rate of inflow in the river (El-Sabh *et al.*, 1997). When this happens, residual flows take sea water upstream, where further evaporation leads to hypersaline water. In the Laguna Madre, Texas, salinity in summer may be twice that in the Gulf of Mexico. In Spencer Gulf, South Australia, salinities reach 39 during summer, and then as the waters cool during autumn they become so much more dense than sea water outside the Gulf that they sink and move seawards (Fig. 7.4). A surface flow of sea water then moves into the Gulf, creating circulation that is the exact opposite of that in a salt wedge.

This categorization of estuaries into specific types is useful, but obscures the fact that many estuaries may show facets of several different patterns. In some Australian estuaries, for example, intense evaporation leads to the formation of a high-salinity 'plug' which effectively prevents mixing of estuarine water with the sea (Wolanski, 1986). In this case, the region above the plug functions as a normal stratified estuary, while the region below functions as a negative estuary. In other cases, where rainfall is strongly seasonal, conditions in negative estuaries fluctuate widely. Thus the Sarada-Varaha estuary in India is negative from April to June, but with the coming of the monsoons river flow may increase sufficiently to turn it into a salt-wedge estuary. The conditions for fauna and flora in negative estuaries are thus often extremely rigorous, and such estuaries are dominated by opportunistic organisms that can move in, feed, and breed rapidly before their populations are decimated.

Patterns of sediment deposition

Estuaries are sediment traps, and indeed as far as biota are concerned, many of the properties of estuaries relate to sediment—where and when it is deposited, how much is suspended, and how mobile it is. Patterns of sedimentation vary with overall estuarine shape and with distribution of salinity, but perhaps the greatest contrast is found between wave-dominated and tide-dominated environments. Where tides are small, the major forces that move sediment are usually waves and river flow. Where tides are large, tidal currents usually outweigh waves in their importance. Thus in a general way, the sedimentary structures in an estuary can be related to tidal amplitude (Hayes, 1975). While there are numerous exceptions to this concept, a split into small tidal range, medium tidal range, and high tidal range estuaries is therefore extremely useful to biologists because from this one value they can predict much that is important about habitat structure.

Microtidal estuaries are those with a tidal range of less than 2 m. They are formed where waves are active in moving sediment, often behind offshore bars—they are thus usually of the bar-built type discussed on p. 136. Because river flow is usually dominant, there is often a salt wedge as the river flows into an area behind the sand bar, and the river may form a delta here (Fig. 7.5). In time this delta may join up with the bar. Depending upon the completeness of the bar, these microtidal estuaries may be called coastal lagoons—such as those on the Gulf coast in Texas or the Gippsland lagoons in southeast Australia.

Fig. 7.5 A comparison of microtidal, mesotidal, and macrotidal estuaries. In microtidal estuaries, wind and wave effects dominate the outer regions, but the inner estuary may have a river delta. In mesotidal estuaries, tidal forces produce substantial flood and ebb deltas near the mouth, but the estuary has extensive tidal flats. In macrotidal estuaries, the mouth is dominated by linear sand bodies, but extensive tidal flats line the estuary's edges. (After Hayes, 1975 and Pethick, 1984.)

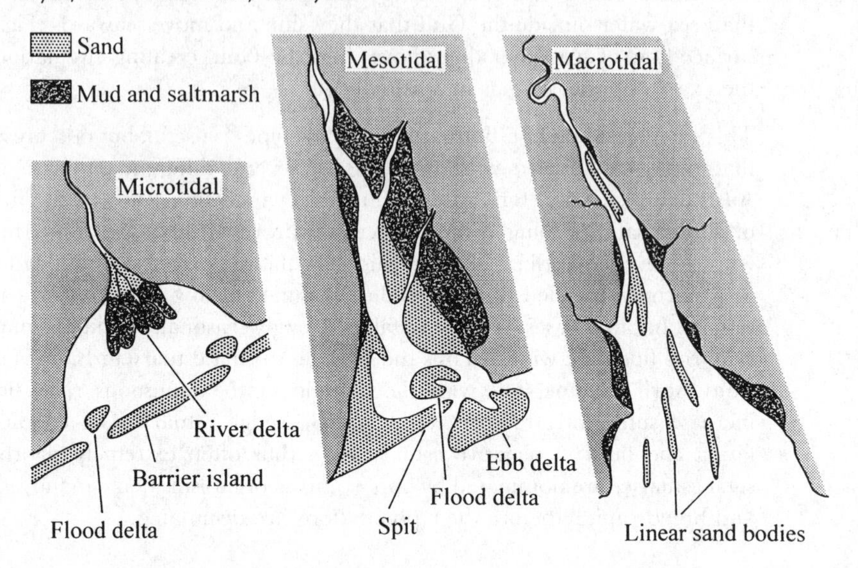

Mesotidal estuaries have a tidal range between 2 and 4 m. Tidal currents as well as waves are important in moving sediment, so while there may be sand bars at the estuary mouth (Fig. 7.5), tidal forces tend to produce deltaic deposits both inside the bars (flood deltas) and outside (ebb deltas). In their upstream reaches, mesotidal estuaries have tidal channels that meander between extensive salt marshes. Many of the estuaries within Chesapeake Bay fall into this category, along with San Francisco Bay and the Thames (UK).

Macrotidal estuaries occur where the tide range is higher than 4 m. They tend to have a wide mouth that contains long, thin bodies of mobile sand, instead of ebb and flood deltas (Fig. 7.5). Overall they are usually funnel-shaped, with fringing mudflats and salt marshes or mangrove swamps, especially towards the head. Because tidal currents are strong, macrotidal estuaries are usually well-mixed in terms of salinity, and they have high turbidity levels. Good examples are the Bay of Fundy (USA/Canada), the Severn and Humber (UK), and the Hwang Ho (China).

Types of habitat

Estuaries and coastal lagoons contain all the habitat types that have so far been discussed in detail—mud, sand, salt marsh, or mangrove, as well as hard substrata. But because of their dynamic nature, discussed above, they also impose on these habitats additional characteristics that affect the fauna and flora. Primary among these is salinity. As we shall see, species numbers in estuaries decline with distance from the sea, and this reduction in numbers is usually attributed to salinity. Because of the importance of the subject, much of Chapter 8 is devoted to salinity, and some of the other characteristics of estuarine and lagoonal habitats are considered here.

Estuarine soft sediments

The establishment of dense or diverse communities depends to some extent upon the existence of a substratum with long-term stability. Estuarine soft sediments may conspicuously lack this stability because of the fierce tidal currents discussed above, and shores in larger estuaries may in addition have to face severe wave action. In the Severn estuary, some intertidal sand banks form 'sand waves' over 1 m in height, but these are so unstable that they face upstream on the flood tide and then reverse direction to face downstream on the ebb (Hawkins and Sebbage, 1972). Not surprisingly, the fauna of offshore sand banks in the Severn is extremely sparse.

Estuarine mudflats may also experience severe erosive forces. In the Humber estuary, populations of the bivalve *Macoma balthica* live primarily in the top 6 cm of mud (Fig. 7.6). Yet the height of the sediment surface varies from month to month by several centimetres, so many individuals are washed out of the sediment (Ratcliffe *et al.* in Jones and Wolff, 1981). Indeed the spat (recently settled juveniles) of *Macoma*, which live in the surface 1 cm, must be transported

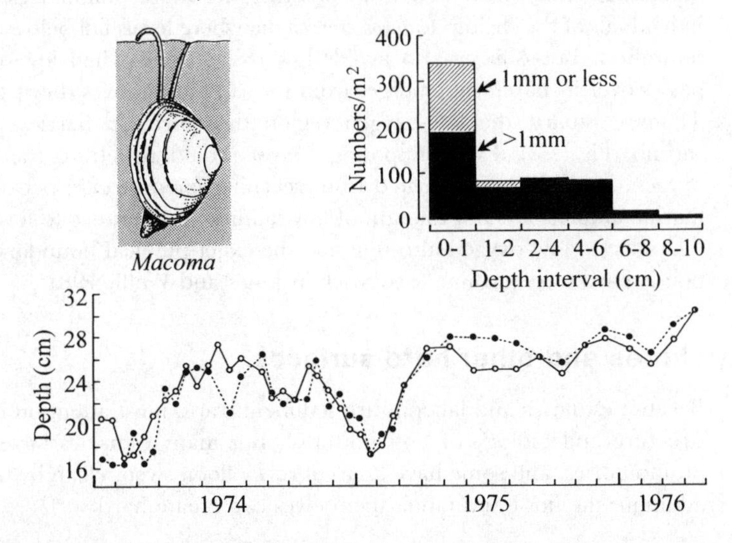

Fig. 7.6 Variation in sediment depth on a tidal mudflat in the Humber estuary (UK), in relation to burrowing depth of *Macoma balthica*. The main graph shows sediment above two plates buried in the mud, over a period of 28 months. Sediment erosion and accretion of several cm occurred rapidly at times. The histogram shows depth distribution of *Macoma* on 20 January 1976. Most spat were in the top 1 cm, and the majority of the population was in the top 6 cm. (After Ratcliffe *et al.* in Jones and Wolff, 1981.)

between different areas very frequently. Measurements of the height of shore stations in the Severn estuary demonstrate that drastic changes in the depth of mudflats can occur in very short times. Here gradual accumulation of over 0.5 m occurred between March and May, followed by erosion of 0.4 m again within a period of 3 weeks (Underwood *et al.* in Jones, 1995). In this case, erosion was mostly caused by wave action.

Instability is thus an extra hazard in macrotidal estuaries. In addition, the inhabitants of sublittoral estuarine sediments face a problem related to the distribution of the turbidity maximum. This problem occurs because, even if there is no stratification imposed by salinity, and the estuary is 'well-mixed', suspended sediment may cause similar effects: the sediment forms density layers which gradually sink through the water column, so that dense layers at the bottom become immobile and do not mix with overlying water. Under these circumstances, which may last for several days (Kirby and Parker, 1983), the biota may use up all the available oxygen, and the resulting anoxia will kill much of the benthos.

Possibly more widespread than these oxygen minima associated with turbidity layers are the regions of low oxygen produced primarily by sewage effluents. Many estuaries experience what is called an 'oxygen sag'—a region where oxygen falls below the 100% saturation normal in the sea and in rivers. In the Thames, for instance, dissolved oxygen fell to less than 5% saturation in some reaches during the 1970s (Fig. 7.7). Despite recent improvements in some countries, low oxygen is still a characteristic of estuaries (Clark, 1997). Very low levels of oxygen can kill the fauna, as we have seen in the sublittoral sediments of Lough Hyne (p. 126). In the Thames and the Clyde, low oxygen restricts the distribution of various oligochaete species, and only the most tolerant can survive in regions where the oxygen sag passes (Hunter in Jones and Wolff, 1981). More mobile animals such as fish tend to avoid low-oxygen regions, and salmon are thus prevented from entering the more polluted estuaries. Most individuals of the shrimp *Crangon crangon* die where levels fall below about 17.5% saturation. But *Crangon* also avoids low oxygen and when low-oxygen water passes over its habitat it emerges from the sand and moves downstream. In the Thames estuary, the low-oxygen region thus forms a barrier to migrating shrimp (Fig. 7.7). Even supposing *Crangon* could penetrate the plug of low oxygen, it would be prevented from reaching far upstream because like most marine animals it is not tolerant of low salinity. As tolerance to low salinity and tolerance to low oxygen also interact, the exact physical boundary to penetration is hard to determine (Sedgwick in Jones and Wolff, 1981).

Rocky shores and other hard surfaces

Because estuaries and lagoons are sediment traps, most attention is paid to the structure and biology of 'soft bottoms'. But many estuaries have some rocky promontories, and some have areas of rocky floor, swept clean by tidal scour. In addition, the flora and fauna themselves can create hard surfaces.

Fig. 7.7 The importance of low-oxygen levels and low salinity in limiting penetration of the Thames estuary by the shrimp *Crangon crangon*. Conditions are shown as recorded in September 1972. Filled circles show salinity, and left-hand hatching shows the region where values are too low for shrimp. Oxygen (open circles) shows a 'sag' in the mid reaches, where levels are two low for *Crangon* survival (right-hand hatching). This low oxygen 'plug' forms a barrier to migrating shrimp. A distance of 20 km marks London Bridge. (After Sedgwick and after Hunter, both in Jones and Wolff, 1981.)

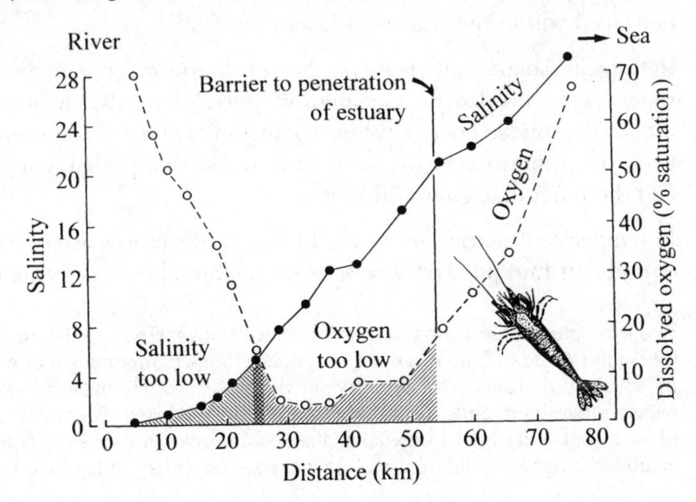

Rocky shores at the mouths of estuaries have communities very similar to those on nearby open coasts (Little and Kitching, 1996). With distance up-estuary, however, there is a progressive decline in species numbers. Partly this is to do with decreasing salinity and increasing salinity variation, as shall be discussed in Chapter 8. But partly the decline is related to two other effects—in some places the rocks are covered with a layer of deposited silt, and in others they are exposed to currents with a high suspended sediment load which acts as efficiently as kitchen scouring pads in abrading the rock surface. We take these two effects in turn.

Marine rocky shores often appear 'clean' except for a growth of macroalgae and populations of encrusting fauna such as barnacles. In fact, the rocks are covered by a thin and often scattered layer of diatoms and cyanobacteria that provide food for mobile grazers such as limpets. Estuarine rocky shores, in contrast, often appear 'dirty' with a thick layer of deposited sediment in which the diatoms and cyanobacteria form a three-dimensional network. In the Severn estuary, for example, the mud layer is held in place by a meshwork of the chain-forming diatom *Melosira nummuloides*. When grazers are excluded from such rocks, the mud-and-diatom layer may persist throughout the year, but when limpets are allowed access they scrape away patches with their radulae and the thickness of the layer declines (Little and Smith, 1980). Silt layers like this, bound to the rock, may inhibit settlement by larvae of animals and spores of plants, so that even without grazing, estuarine rocks may stay bare of macroflora and fauna for years.

In the sheltered waters of coastal lagoons, such layers of associated silt and algae, termed 'Aufwuchs', can be common. Within the layer a complex ecosystem of meiofauna may develop. In the Swanpool, a brackish-water lagoon in southwest Britain, the small oligochaete *Nais elinguis* fluctuates in numbers but may reach up to 200 000/m^2. *Nais* feeds on sediment particles and on the surface layer of diatoms, while other meiofauna such as nematodes and copepods feed within the Aufwuchs layer (Fig. 7.8).

In tropical estuaries and lagoons, the only hard surfaces may be provided by the roots, stems and leaves of mangrove trees. Here also, a film of sediment is usually deposited. There have been few studies of the constitution of this film, or who eats it, but mangrove snails such as *Littoraria* probably utilize the detritus and the fungi that grow within it.

In temperate estuaries and coastal bays, hard surfaces are provided by beds of mussels. In Europe, *Mytilus edulis* can tolerate life on mudflats despite being a

Fig. 7.8 Two examples of hard substrata with sediment cover. The top diagram shows a mussel bed in the Wadden Sea, where *Mytilus edulis* forms biodeposits that prevent settlement by sessile organisms. The alga *Fucus vesiculosus* forma *mytili* is held in place by the mussel's byssus threads. (After Albrecht, 1998.) The lower diagram shows the thin layer of sediment and algae (Aufwuchs) that coats rocks in a coastal lagoon. The layer is inhabited by several different meiofaunal species. (After Little, 1986.)

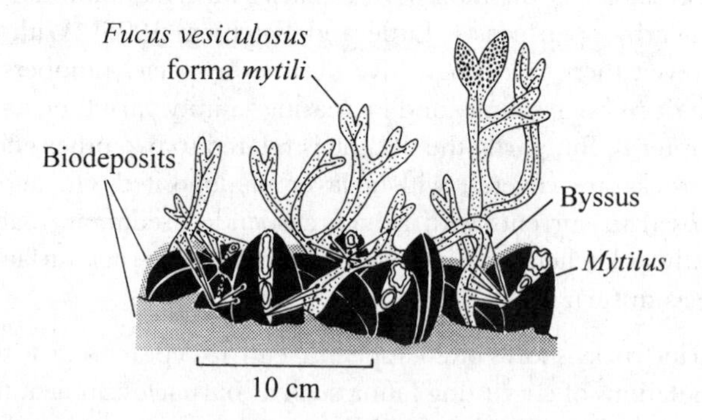

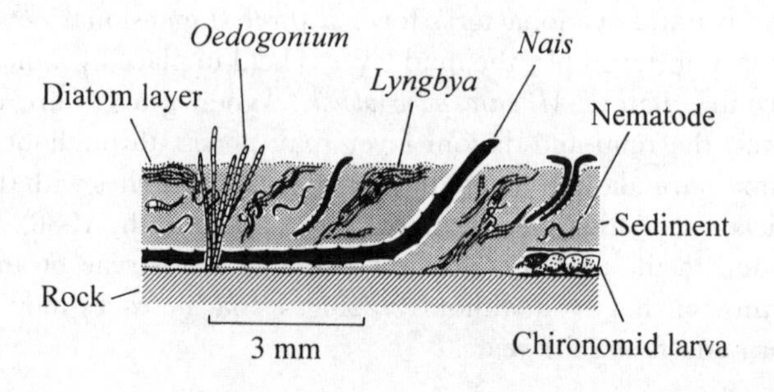

suspension feeder. The Exe estuary, for example, contains beds of mussels that can be as large as 14 ha. In the Wadden Sea, Germany, extensive mussel beds cover the sediment, but seldom show growth of the macroalgae that are found on other hard surfaces such as docks and pilings. They may have a dense cover of the alga *Fucus vesiculosus*, but this is a form which has no holdfast, is kept in place by the byssus threads of the mussels, and does not reproduce sexually (Fig. 7.8). Why are the mussels not colonized by other seaweeds? Experiments using cages to exclude grazing snails, and dead mortar-filled shells to mimic mussels, show that both grazing and the activities of the mussels are important (Albrecht, 1998). First, when the grazing periwinkle *Littorina littorea* is excluded, ephemeral red and green algae such as *Porphyra* and *Enteromorpha* do colonize, showing that it is the grazers that prevent their growth. But under these circumstances, *Fucus* sporelings do not survive. When the experiment is repeated with dead mussels, *Fucus* sporelings show better survival—so it seems that the mussels themselves prevent sporeling development. What the live mussels actually do is to produce enormous quantities of 'biodeposits'—faeces and pseudofaeces (i.e. rejected particles bound up in mucus)—which accumulate on the shells. Despite the fact that mussels also clean parts of their shells with the foot, this accumulation effectively covers settling sporelings and stops them from growing. In this example, *Mytilus* provides a hard surface, but the action of grazers and of the mussels then prevents its use except by one form of alga.

The opposite problem for colonizing benthos is that of scour, but this has, surprisingly, been little studied. In macrotidal estuaries the effects of scour can be extreme. For example, a bed of the alga *Ascophyllum nodosum* on a steep shore in the Severn estuary almost disappeared one year, probably ripped off by tidal scour and wave action, and the cliff then remained without significant algal cover for several years (Little and Kitching, 1996). On many shores in the Severn, macroalgae are absent from the low shore, and this absence is probably due to high silt levels in combination with tidal scour, which prevent successful settlement of sporelings (Little and Smith, 1980). In the sublittoral, scour problems are often much worse. Thus the majority of the bed at the mouth of the Severn estuary has no accumulated sediment and has a sparse community dominated by the tube-dwelling polychaete *Sabellaria* (Warwick and Davies, 1977). Here the rocks and boulders are worn smooth by rolling in very strong currents, and abraded by mobile gravels. This type of sublittoral habitat contrasts strongly with that of estuaries with lower tidal currents, the mesotidal and microtidal forms.

Lagoonal habitats and consequences for their inhabitants

Because lagoons are relatively enclosed and sheltered, in comparison with estuarine and coastal marine environments, they tend to accumulate sediments which despite being very fine and flocculent, remain stable. These sediments fill much of the lagoon basins, and lagoons are consequently shallow—few are deeper than 10 m, and many are less than 1 m. In all other respects, however, lagoons present their inhabitants with very changeable habitats, so much so that some lagoonal experts have concluded, in despair, that every lagoon is unique (Colombo in Barnes, 1977). In the long term, some variation occurs because lagoons occupy an intermediate position in evolutionary sequences from coastal bay to freshwater pond—or from coastal bay back to coastal bay if the enclosing barrier is eroded. In the short term, variability may occur in many characteristics—in salinity, in nutrient concentration, in oxygen content, or in temperature, for example. Changes in these variables are not usually due to tidal influence, as they are in estuaries, so the time scale of change is longer than 12.4 h. But mixing of bottom saline water with surface water can increase salinity very suddenly, and oxygen concentrations can change drastically over a 24 h period (Barnes, 1980). More widespread than these sudden changes are seasonal effects. For example, in the Commachio lagoon, Italy, temperature varies from a winter low of 8.5°C to a summer high of 29°C, salinity varies from 26 to 48, and pH from 7.7 in winter to 9.5 in summer (Colombo in Barnes, 1977). In the tropics, even more drastic variation in some factors can be seen, as in the Caimanero lagoon, Mexico (Fig. 7.9). Here evaporation can increase salinity by as much as 2 units in 1 day, although the overall hydrological situation is quite complex: during the wet season, more water flows into the lagoons from the rivers, but more salt water enters from the sea because of an annual cycle of sea-level variation. Overall, salinity thus declines in the wet season in Caimanero, but does not fall below about 8 (Edwards, 1978).

Fig. 7.9 Annual changes in Caimanero lagoon, Mexico. The lower graph shows mean air and water temperatures. The upper graph shows average monthly salinities in the lagoon and in the canals feeding it. *Penaeus* spp. enter as postlarvae in the wet season (low salinity and high temperature) and grow into adults that leave in the dry season (rising salinity, lower temperature). (After Edwards, 1978.)

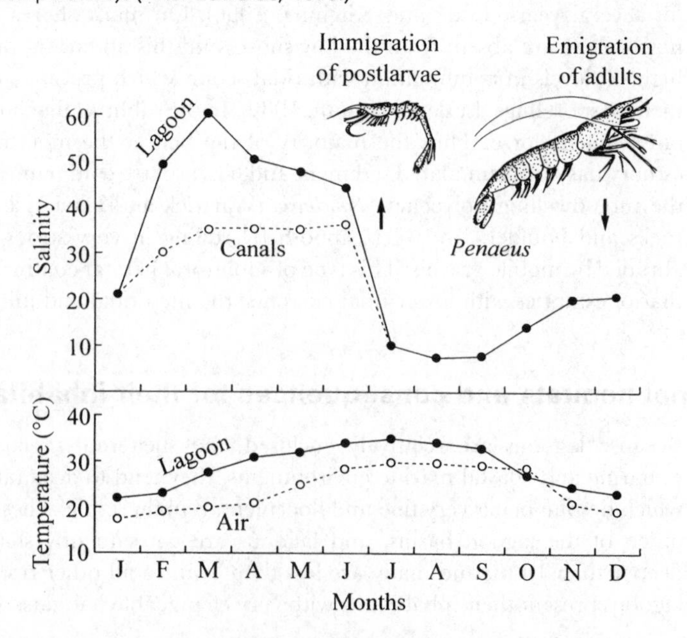

Lagoons often have high concentrations of nutrients such as nitrate, phosphate, and silicate, because new supplies constantly arrive in the rivers and streams that flow into them. These nutrients may be recycled from the sediment to the water column if the water is shallow enough for mixing to occur all the year round. But in the Commachio lagoon, parts of which are 10 m deep, nitrate and silicate are considerably depleted in summer, as these nutrients precipitate in dying phytoplankton. When this depletion occurs, phytoplankton growth rates fall. In the Swanpool, southwest England, even depths of 2.5 m allow phosphate to diminish to levels that could limit phytoplankton growth, because the reservoir of phosphate becomes trapped in dense saline water at the bottom of the pool (Crawford *et al.*, 1979). For the most part, however, lagoons retain high nitrate and phosphate levels, often allowing extremely high rates of primary production. Thus in the Caimanero lagoon, Mexico, diatoms in the water column may fix more than 2 gC/m^2 per day, roughly 10 times the rate found in neighbouring seas. Added to this, the submerged pondweed *Ruppia* fixes large amounts of carbon, and with mangrove detritus brought in through the channels linking the lagoons, the overall supply of detrital carbon is enormous.

In Caimanero, as in many other lagoons, there are two types of fauna that depend upon this detrital supply and the warm, sheltered conditions. First are the resident species, which complete their life cycles within the lagoon. One of the most conspicuous of these is a snail, *Cerithidea* (Fig. 4.1), which reaches densities of $500/m^2$, or 30 g dry flesh weight, and dominates the epifaunal biomass. In temperate lagoons, a variety of other snails such as *Hydrobia* spp. (Fig. 4.7) crop the benthic diatom flora and consume the fine detritus, in company with crustaceans such as amphipods, polychaete worms, and others (Barnes, 1980). While species numbers in lagoons are usually low, the large detritus supply allows dense populations and high rates of secondary production. The species present vary very much from one locality to another, and lagoons that have no channels connecting them to the sea have rather random combinations of species reflecting the importance of chance factors in affecting colonization (Barnes, 1988).

Some of the species common in lagoons are closely related to species that occur in neighbouring coasts and estuaries. These lagoonal species have presumably evolved in response to the characteristics of non-tidal coastal waters, and they thus emphasize some of the lagoonal features that are biologically important. In Europe, the common cockle of estuarine flats, *Cerastoderma edule*, is replaced in lagoons by *C. glaucum*. The lagoonal species has a very wide salinity tolerance and can withstand very low oxygen tensions. It does not always lie buried in the mud like its mudflat relative, but when juvenile climbs actively on submerged macrophytes. The two species have similar breeding patterns, but the larvae of *C. glaucum* live in the plankton for only a week, whereas those of *C. edule* live for 2–6 weeks. Altogether, the two species are extremely similar except for the increased tolerance of *C. glaucum*, and have evidently not diverged much from the parent stock.

Also in Europe, the hydrobiid snails contain both estuarine and lagoonal species. The commonest natives are *Hydrobia ulvae*, found on estuarine mudflats, and *H. ventrosa*, which lives in lagoons (see p. 73). These two differ in salinity tolerance, in the size-fractions of detrital food that they prefer, but most notably in their reproductive specializations. Thus *H. ulvae* produces many veliger larvae which, although they remain near the substratum, allow the species to be widely dispersed in estuaries; while *H. ventrosa* produces fewer eggs that hatch directly into young snails. Suppression of larval stages and concentration upon more of a K-strategy lifestyle (i.e. maximizing competitive ability) thus seems to be typical of lagoonal species. In comparison, estuarine species have more *r*-strategy attributes, which maximize mechanisms that aid dispersal and rapid reproduction. This may seem surprising, as we have just been emphasizing the variability of conditions within lagoons. However, in terms of biological response, lagoonal variability is evidently slow enough to allow tolerance to evolve as a mechanism to cope with it. In estuaries, the variability on a tidal (12.4 h) timescale means that individuals may die whatever their response—from desiccation, from salinity variation, or from predation either by wading birds or migratory fish. In this case, *r*-strategies make sense.

The second type of fauna found in lagoons consists of migratory species. The most widespread of these are prawns, but crabs and fish also use lagoons as 'nursery' areas in which the young grow up before recruiting to the adult population at sea. The migrations of these temporary species can lead to enormous variations in the populations within lagoons. In the Swanpool, *Palaemonetes elegans* reaches high summer population densities and the average weight of individual prawns may increase from 0.7 mg in June to 30 mg in September. In Caimanero, *Penaeus* spp. migrate in as postlarvae in the wet season, then grow over a period of 6 months into adults that migrate out again as salinity begins to rise in the dry season (Fig. 7.9). In the high temperatures of this lagoon, prawns may grow as fast as 1.5 mm each day. Not surprisingly, a flourishing prawn fishery crops some of this production, catching maybe 1400 tonnes/year, and many other tropical lagoon systems are the basis of fish and crustacean harvests.

Techniques

It is not the purpose of this book to detail methods of physical or chemical survey and measurement. Excellent accounts of how to investigate estuarine hydrography and sedimentation are given by Dyer (1979, 1997), and of how to investigate chemical aspects of estuaries by Head (1985). Portable meters for measuring most variables, from oxygen and salinity to turbidity and nutrient concentrations, are now widely available. In terms of surveying the variety of habitats within estuaries—and on other coasts—one development that shows promise is remote sensing, and this may soon become a realistic option (Sotheran *et al.*, 1997).

8 The estuarine benthos and its distribution

Because benthic habitats in estuaries vary from solid rock to sand and mud, and from bare regions to vegetated beds of salt marsh, seagrasses, or mangroves, they are homes to a wide variety of inhabitants, especially in parts of the tropics. But in comparison with the sea and fresh water, their species diversity is in fact surprisingly low, and there are some areas of estuaries that have very few species at all. One of the major preoccupations of estuarine biologists has been to explain the distribution of diversity in estuaries. Are physical forces so strong that habitats are physically determined, as most people believe to be so on exposed sandy beaches (Chapter 3)? Is salinity the overriding control factor? Or are there more subtle interactions between a variety of physical variables and the organisms themselves? We begin by describing some patterns of species distribution within estuaries before considering controlling factors.

Species distribution in estuaries

Early estuarine studies showed that marine species die out with distance up-estuary, and that they are replaced by freshwater species towards the riverine end. For example, in the Tay estuary, Scotland, marine species die out over a distance of 30 km from the sea, and are gradually replaced by freshwater species, so there is an area about 25 km from the sea where there is a minimum number of overall species (McLusky, 1989). This picture, with a minimum at some point, seems consistent for most estuaries, and is also found in tideless seas such as the Baltic.

Detailed studies of the decline in numbers of intertidal marine species, with distance from the sea, have now been carried out in many temperate estuaries. In the Severn estuary, changes in numbers of rocky-shore and soft-shore species can be compared by simple linear plots (Fig. 8.1). Here there is a rapid decline upstream on rocky shores, but a much smaller decline on soft shores. There is a great deal of variation between sites, suggesting that the idea of some kind of smooth progression upstream may be misleading: individual site characteristics are very important.

Equivalent surveys examining subtidal areas have disclosed the extremely patchy nature of bottom faunas in estuaries. For example, a plot of species

Fig. 8.1 A comparison between numbers of rocky-shore and soft-shore animal species in the intertidal zone of the Severn estuary. Numbers are plotted against distance towards the sea from the limit of tidal penetration. (After Boyden and Little, 1973.)

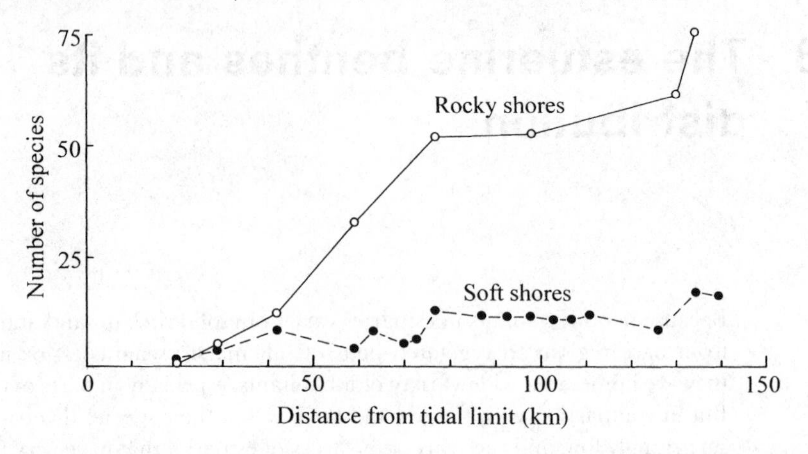

richness in the Forth estuary, Scotland, shows enormous variation between sampling sites until at the very head of the estuary, species numbers drop dramatically (Fig. 8.2). Nevertheless, the fauna can be divided into a series of 'communities', of the types discussed in Chapter 6 (Elliott and Kingston, 1987). Many of the communities in the outer regions are related to sediment type, water currents, and depth, while in the inner estuarine areas the very low diversities correlate with low salinity.

Fig. 8.2 The distribution of species numbers in the sublittoral infauna of the Forth estuary and Firth of Forth, Scotland. Circles show means and bars show range. Above 30 km from the tidal limit, only single samples were taken. (After Elliott and Kingston, 1987.)

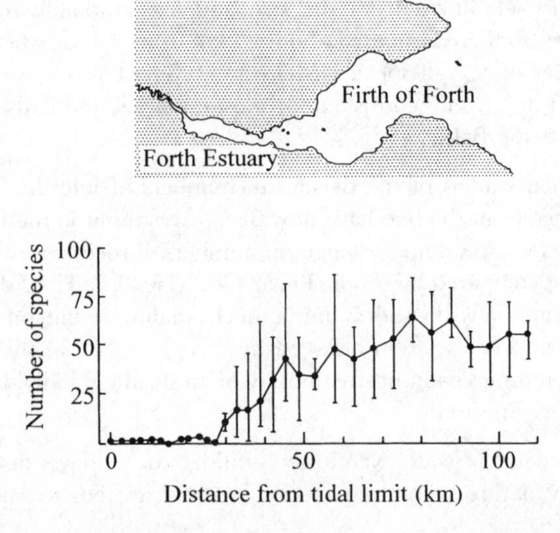

The distributions of subtidal benthic communities in the Chesapeake Bay and in the Brisbane River estuary, Australia, follow similar patterns (Boesch in Coull, 1977). Near the mouths of both these estuaries, communities change rapidly as the outer coarse sands give way to estuarine muds and muddy sands. Over larger distances upstream, stenohaline marine species (those intolerant of low salinities) give way to more euryhaline (tolerant) forms, which can live in a range of salinities (Fig. 8.3). In the Brisbane River, these upstream changes may relate to seasonal decreases in salinity during high river discharge, but the same does not apply in the Chesapeake, where salinity is fairly constant round the year. Towards the upper end of both estuaries, 'estuarine endemic' species occur—those that are not normally found downstream. Finally, at the riverine end, freshwater species appear. The pattern is certainly one of increasing tolerance of low salinity upstream, but it should be noted that most of the species found (except the freshwater ones) can probably withstand full-strength sea water, so salinity can hardly explain distribution downstream.

Fig. 8.3 The distribution of the sublittoral infauna in the Brisbane River estuary, Australia. Histograms show species groups or 'communities'. Symbols within the histograms show how common each species group is at each station. Thickness of the histograms is in proportion to the number of species in each group: group G has one species, group D has seven species. (After Boesch in Coull, 1977.)

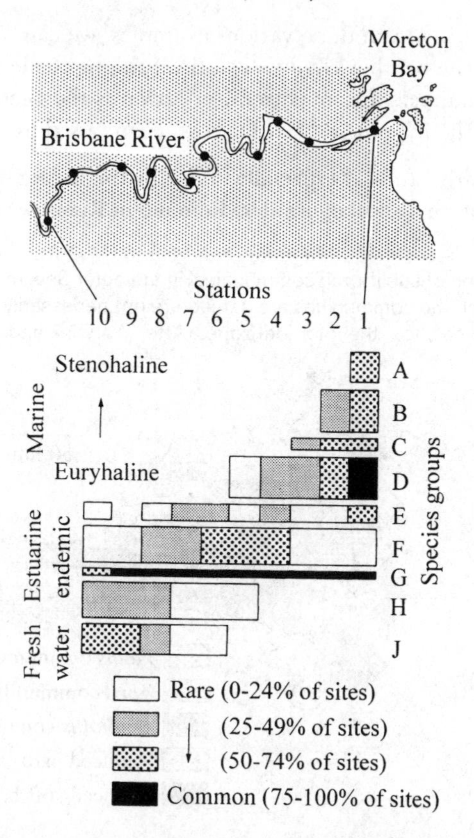

A final example is provided by the distribution of subtidal benthos in the Severn (Fig. 8.4). Here the patchy nature of communities is well seen, but there is also a progression of different communities upstream. At the mouth, where the substratum is mostly sand, the fauna is dominated by polychaetes and bivalves—the so-called *Venus* and *Abra* communities, although the bivalve *Venus* itself is actually rare. Upstream of these is a wide swathe of a very mixed community containing over 200 species. This is called the *Modiolus* community, although again, the mussel *Modiolus* itself is rare. Further upstream again is an area of hard substratum which supports only about 20 species—the 'reduced hard-bottom' community; and towards the head of the estuary this is replaced by a 'reduced soft-bottom' community containing only a few oligochaetes and polychaetes. The influence of substratum is very apparent in this example, and shall be returned to later.

The situation in tropical estuaries has not, as yet, been investigated in such detail. Species diversity in some tropical estuaries is lower than in temperate ones because of the drastic seasonal fluctuations in the environment. On the Indian coast, for example, seasonal monsoons create violent changes in salinity, sediments, dissolved oxygen, and nutrients (Alongi, 1990). In tropical estuaries without such fluctuations, diversity may be high but the distribution of species requires further study.

Summarizing from these various examples, we can say that the species composition of the benthos changes drastically along the length of most estuaries, and species numbers are much reduced towards the head. We now need to discuss some of the hypotheses that have been proposed to explain these observations.

While early studies, especially those in the Baltic, often placed emphasis on salinity as an overall controller (Remane in Remane and Schlieper, 1971), it was

Fig. 8.4 Distribution of sublittoral 'communities' in the outer Severn estuary. Note that the genera after which the communities are named are not necessarily common in the areas shown. Dotted lines show the intertidal zone. (After Warwick and Davies, 1977.)

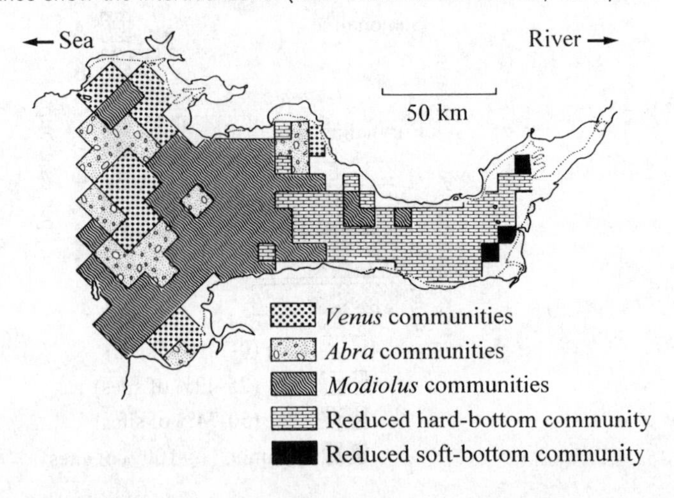

also remarked that areas of salinity between sea water and fresh water—'brackish waters'—have many other properties that are challenging for the biota. For example, stagnant zones occur, in which oxygen may become depleted, and hydrogen sulphide may then accumulate. There is a range of substratum types. Salinity fluctuations may also be more important than absolute salinity values. To these points may be added the direct effects of current speed or 'bed stress', and the host of 'biological' influences such as competition, predation, and food supply. We begin with salinity, however, because it has for so long been regarded as dominant.

Salinity as a controlling factor

Sea water and brackish water

Sea water is a chemical broth dominated by the ions sodium and chloride, but with substantial fractions contributed by other inorganic ions (Table 8.1). There are also low concentrations of many other inorganic ions and organic substances, but these make up only about 0.1 g/kg—less than 0.5% of the total. This total, in average sea water, makes up 35 g/kg, now known as a salinity of 35 (see p. 169). The figure of 35 is, however, far from universal (Summerhayes and Thorpe, 1996). Most of the Pacific has salinities between 34 and 36, but in the centres of the North and South Atlantic, salinity rises to 37, while in high latitudes of all oceans it falls to about 33. A symposium convened in Venice in 1958 specifically to discuss salinity concluded that anywhere between 30 and 40 could be considered 'euhaline' (Anon, 1959).

Where sea water is diluted with river water to become brackish, the Venice system uses the term 'mixohaline'; i.e. for salinities between 30 and 0.5. Values below 0.5 are considered fresh water. The Venice system also defines many intermediate categories of the mixohaline zone, but the only one in common usage is 'oligohaline', referring to salinities between 5 and 0.5. At these low salinities, the composition of the water may be significantly affected by the ion content of river water, so there is often relatively more calcium and bicarbonate, and less chloride, than one would expect from dilution with, say, distilled water (Head, 1985).

Table 8.1 Major constituents of sea water (salinity 35). (After Sverdrup *et al.*, 1942 and Harvey, 1963)

	Concentration (g/kg)	Percentage of total weight
Chloride	19.4	55.0
Sodium	10.8	30.6
Sulphate	2.7	7.7
Magnesium	1.3	3.7
Calcium	0.4	1.2
Potassium	0.4	1.1
Bicarbonate/carbonate	0.1	0.4

Responses of organisms to brackish water

It has been evident in several previous chapters that many biologists regard salinity as an extremely important variable for many animals and plants. For example, salt marsh and mangrove plants are limited to saline areas but secrete salts from their tissues to maintain viable internal salt concentrations. Salt marsh snails become bloated when covered by low-salinity water. In mudflats, species diversity of infauna decreases as salinity declines. Why should salinity be such a powerful influence on marine animals? There are many levels on which we could discuss this question, but it may be best to consider it first from the physiological angle.

Most marine invertebrates have blood which is very similar in composition and in overall concentration to sea water. They are unable to control this composition in the face of changes outside and are thus called osmoconformers: when total osmotic pressure of the water (which is equivalent to its salinity) changes, that of the blood changes too. A large number of species cannot tolerate much decrease in internal osmotic pressure—they are called 'stenohaline' and reduction beyond a salinity of approximately 30 often results in death. Stenohaline forms are therefore absent from most estuaries and lagoons because the drop in salinity forms an absolute barrier.

Species that are able to tolerate large overall reductions in salinity are called 'euryhaline', so these are the ones that dominate in brackish waters. Most are unable to control their internal composition—they are osmoconformers too—but unlike the stenohaline species they have physiological mechanisms that protect their cells against the effects of dilution. A minority of euryhaline species can osmoregulate—particularly some crabs and shrimps, and many fish—so they are able to maintain the osmotic pressure of their blood and tissues higher (and in some cases lower) than that of the sea water outside. However, the ability to osmoregulate in brackish-water animals is rare: most species have developed tolerance to a fine art instead (Little, 1990).

Does the degree of salinity tolerance shown by estuarine plants and animals determine the distance they penetrate into brackish water? Some authors are convinced this is so. For instance, a graph produced by Remane (in Remane and Schlieper, 1971) plots distribution not against distance into estuaries, but against salinity (Fig. 8.6). Many others have followed this tendency, and indeed, because salinity is probably the most easily measured variable in estuaries, and a good measure of position within an estuary, many estuarine workers refer their studies of biota to salinity. In a recent study in North America, for example, zones of salinity within estuaries have been related to species distributions (Bulger *et al.*, 1993). Nevertheless, as we have already emphasized, changes in salinity are usually accompanied by changes in substratum, current flow, sediment stability, oxygen levels, food supply, and so on. These 'confounding variables' make it very difficult to decide whether salinity itself really has an overriding effect, or whether it is merely 'one of the variables'.

The fact that stenohaline species cannot tolerate low salinities makes it very likely that salinity is a prime reason for their exclusion from estuaries, though there is no proof that it controls their exact distribution—few species are found at their physiological tolerance limits in nature. For euryhaline species, it is perhaps most likely that salinity defines a range of habitats within which they can grow—and we can regard this as one definition of their potential niche. Other factors then almost certainly reduce this to a 'realized niche': as with stenohaline species, it is unlikely that euryhaline species will be distributed over the full salinity regime they can tolerate in, say, laboratory conditions.

Supporting evidence for a direct effect of reduced salinity on penetration into brackish waters comes from the observation that many species of animals and plants are reduced in size in dilute conditions. This effect has been recorded particularly in the Baltic, where the mussel *Mytilus*, the clam *Mya* and the cockle *Cerastoderma* all diminish in size along a salinity gradient (Remane in Remane and Schlieper, 1971), reaching minimum lengths in a salinity of 5 that are less than half those achieved in the sea. Among the algae, *Laminaria saccharina* shrinks to a few centimetres in length in the Baltic, in comparison with its length in the sea of over a metre. However, the exact cause of this dwarfism has not been determined. While it may be partly due to some direct effect of salinity, it may also involve inadequate food supplies or nutrients or interactions with other organisms, as shown by observations on the lugworm *Arenicola marina* (Mettam, 1980). In the Severn, this species diminishes in size up-estuary (Fig. 8.5). Part of the reason for the small size of the upstream populations is that they do not reproduce—they arise from juveniles or larvae that float upstream and then

Fig. 8.5 Size and breeding condition of the lugworm *Arenicola marina* in relation to distance up the Severn estuary. Overall length decreases and percentage breeding also decreases. (After Mettam, 1980.)

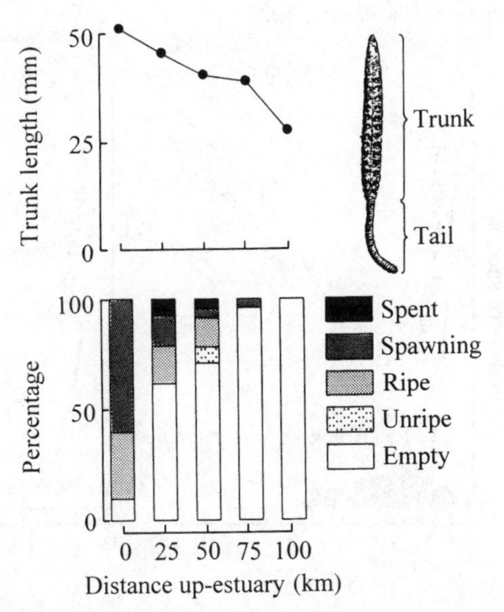

grow during the summer, but die out in winter. Individuals in upstream populations may therefore be small because they are young, *not* because they are dwarfed by some physical effect of salinity.

Are there distinct brackish-water species?

Most of the benthic animals and plants in estuaries are euryhaline species of marine origin, except in regions with a salinity below 5 (the oligohaline zone). Here animals of freshwater origin are usually evident, and particularly in lagoonal situations various genera of insects and oligochaetes may be abundant. Are there, in addition, species that are adapted specifically to the intermediate salinities of brackish water? Some of the dominant species in estuaries and lagoons are traditionally referred to as 'brackish-water species'; for example, the polychaete *Nereis diversicolor*, the snail *Hydrobia ulvae*, and the bivalve *Macoma balthica*. Opposing points of view suggest that these are truly brackish-water specialists—or that they are really marine opportunists, and that true brackish-water species are so rare as to be insignificant.

Remane (in Remane and Schlieper, 1971) argued that between salinities of about 3 and 18, a brackish-water biota exists that is only occasionally found in the sea or in fresh water (Fig. 8.6a). At salinities of 5–8, there are very few species at all, and here the brackish-water species constitute a very high percentage of the total. The curve that Remane plotted is a generalized, partly hypothetical one, based on many observations in the Baltic and the neighbouring North Sea, but interpreted subjectively. How nearly does it represent reality?

Fig. 8.6 Two hypotheses relating numbers of species to salinity. In (a) a brackish-water component is drawn. In (b) this is replaced by small increases in the freshwater and marine components. (After Barnes, 1989.)

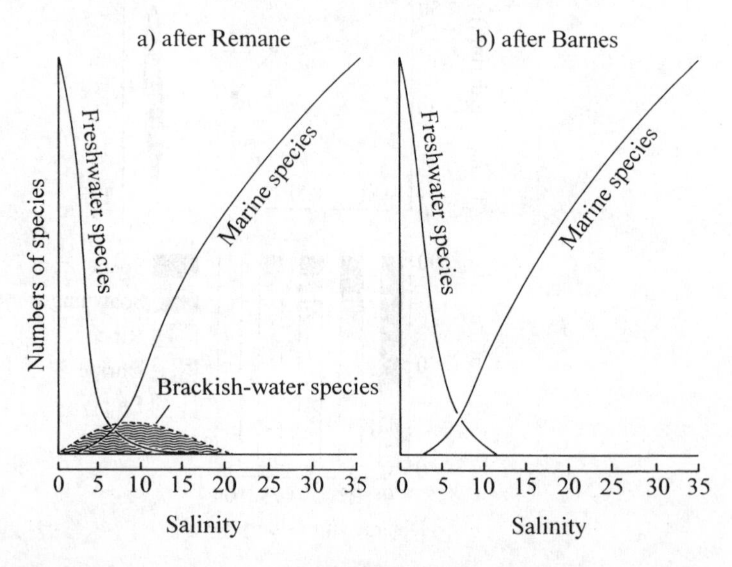

Barnes (1989) discussed this curve and pointed out that many of the species included by Remane in his brackish-water group are in fact often found over a wide range of salinities from 5 to 35. For example, of the mudflat species discussed in Chapter 5, this applies to *Hydrobia ulvae*, *Macoma balthica*, and another bivalve *Mya arenaria*, as well as to many others. *Hydrobia ulvae* has even been recorded 130 m down in offshore marine sediments. In addition, the distribution of many of the so-called brackish-water species along the length of estuaries varies greatly between estuaries (Boyden and Little, 1973), suggesting that if there is a salinity limitation, other factors often override it. Barnes therefore concluded that physiological reaction to dilute brackish water is not the major determinant of the distribution of species in estuaries. In fact, organisms found in estuaries are so exceedingly tolerant of a very wide range of salinities that they could survive along the entire length of most estuaries if salinity were the only controlling factor. In most cases, as we shall see, other factors are probably just as important. Barnes therefore re-drew Remane's original graph, eliminating the category of brackish-water species (Fig. 8.6b). These species may indeed be called 'estuarine', but they do not colonize areas primarily as a reaction to brackish water.

Does the same apply to species in coastal lagoons? Lagoonal communities, like those of estuaries, are dominated by organisms of marine origin (Barnes in Kjerfve, 1994). At a gross level, salinity may determine community composition in lagoons: when the faunas of lagoons in East Anglia, UK, are compared, variation in species richness correlates with salinity (Barnes, 1989). Salinity accounts for only 21% of the variation in species richness, however, despite its wide variation (from 1 to >30), so other factors such as problems of dispersal and colonization are probably more important.

The importance of changes in salinity

Nevertheless, salinity may have important effects on the lifestyle of estuarine animals, especially when it changes. Many infaunal animals retreat into the substratum when salinity suddenly falls or rises. This reaction protects them from sudden change because there is a considerable time-lag in salinity change a few centimetres down in mud or sand. Alternatively, animals that possess them close their shells with similar effect. In both cases, the body fluids can then reach a new equilibrium with the water outside over a protracted period. For animals that cannot move, and for plants, changing salinity is a more major challenge. Possibly for this reason, marine euryhaline species penetrate into regions of much lower salinity where salinity is stable than where it fluctuates. In the Baltic, for example, marine species such as the alga *Fucus vesiculosus* and the isopods *Idotea* spp. are found down to salinities of 5 or 6. In most European estuaries, these species are seldom found where salinity falls below 10. We should note, however, that there are many differences between the Baltic and tidal estuaries other than salinity fluctuations. For example, the Baltic has no tidal mixing, so the suspended sediment load is much lower, and current speeds are less. Both have consequences for the benthos, especially suspension feeders and algae.

There is a voluminous literature on the responses of fauna and flora to sudden changes in salinity (Kinne, 1971). Laboratory experiments show that sinusoidal changes—such as might occur over a tidal cycle—are much less harmful to permeable animals than abrupt ones. Scallop larvae (*Pecten maximus*), for example, do not suffer high mortalities when salinity changes gradually because the tissues are allowed time to acclimate (Davenport *et al.*, 1975). In those animals that have protective shells, into which they can withdraw, it is not surprising to find sensitive mechanisms for detecting salinity change. The snail *Hydrobia ulvae* can detect differences of salinity of less than 10, while another snail, *Potamopyrgus jenkinsi*, can detect salinity differences of as little as 2.8 (Blandford and Little, 1983). Many other molluscs probably use sensory tentacles to warn them of such changes so that they can close the shell or burrow into the substratum.

Effects of substratum, water movements, and water quality

In Chapter 2, we discussed the effects that the particle size composition of sediments may have upon distribution of the benthos. For example, pure sands have a different fauna from fine muds—but it is not at all certain how much of this difference is due directly to differences in grain size, and how much to effects of water flow, which in turn influence such variables as oxygen supply, nutrients, food, and larval settlement. To what extent do sediments and their associated current regimes affect the overall distribution of benthos in estuaries?

While many early studies linked distribution to salinity, there were others that recognized the primary importance of the sediments within which the benthos lives. In the Tamar estuary, UK, the length of tidal immersion and the velocities of tidal and river currents were considered important, and in particular, increased water movements were shown to limit some populations to low tidal levels. In later studies of the Chesapeake Bay and of the Brisbane River estuary (Australia), distributions were mainly related to a salinity gradient as we have discussed above. But it was also noted that at the mouth of both estuaries, sediments changed from mud to sand and gravel: at these stations the largest changes in community structure occur (Boesch in Coull, 1977).

We have already described the distribution of sublittoral fauna in the outer Severn estuary (Fig. 8.3). Does this distribution relate to decreasing salinity upstream, or to changing sediment type associated with changing bed stress? When community type is plotted against bed stress (or average current speed), a close correlation is observed (Fig. 8.7): each community has the centre of its distribution at a particular value of bed stress. In addition, when measured bed stress is used to draw a map predicting distribution of fauna, the predicted

Fig. 8.7 The relationship between benthic community type and bed stress/maximum current speed in the outer Severn estuary. Points show means and bars show standard errors. At the top are shown some of the species common in each community. (After Warwick and Uncles, 1980.)

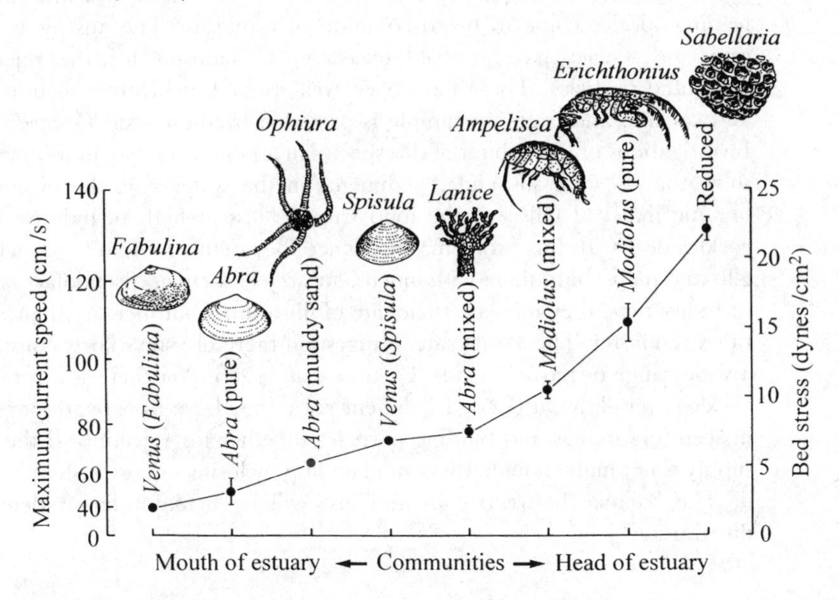

distribution is close to that in Fig. 8.3 (Warwick and Uncles, 1980). In this case there is a strong presumption that bed stress plays a large part in determining substratum character and consequent community type.

This is not to say that the type of substratum is important for all types of organism in all estuaries: it is merely one of a suite of factors. Thus in two estuaries in South Africa, diatom distributions are not related to sediment type, even though sediments vary from fine mud to sand (Watt, 1998). Composition of benthic diatom communities is, on the other hand, strongly dependent upon nutrient status: the eutrophic estuary is dominated by large numbers of small diatoms such as *Navicula* and *Nitzschia* spp. , while the oligotrophic estuary is dominated by larger genera.

One other facet of water quality closely linked with substratum type and water currents is turbidity. In Chapter 9 we consider how it affects life in the water column. It has already (p. 30) been noted how important suspended sediment can be for the benthos when it falls to the bottom as a rain of material. It therefore has an indirect effect on the benthos by interfering with feeding and respiratory mechanisms. For instance, the well known paucity of suspension feeders such as tunicates, hydroids, and sponges in many estuaries may be due to the clogging of the feeding apparatus with silt. Turbid estuaries such as the Bay of Fundy and the Severn have very few of these suspension feeding species. In contrast, 'clean' estuaries such as the Rance (in northwest France) have a dense coverage of bryozoans, tunicates, and sponges in the rocky sublittoral.

Suspension feeders probably have distributions that correlate negatively with distribution of the turbidity maximum, but this has not been demonstrated.

If turbidity acts so negatively on suspension feeders, how do some suspension feeding species come to be so common in estuaries? The answer is that the estuarine species have evolved excellent mechanisms for the rejection of unwanted particles. These have been well studied in bivalves such as mussels and cockles, and a good example is provided by the cockle *Cerastoderma edule.* Investigations of the habitat of this species in an estuarine bay in western France show that as total suspended sediments in the water rise, the proportion of organic material falls—so the food value is less at high turbidities. How do cockles deal with this problem? In essence they detect inorganic particles using gills and palps, bind them with mucus, and reject them as 'pseudofaeces'. As the turbidity rises, they increase their rate of filtration—but increase their rejection rate even faster. The actual rate of ingestion therefore stays fairly constant over a wide range of particle loads (Urrutia *et al.*, 1996). Yet there is a problem for cockles overall: as suspended sediment rises, they have to expend more energy in secreting mucus and binding particles together for rejection. If the organic supply is not high enough, they could end up by losing energy instead of gaining it. The balance between gain and loss will be a major factor determining distribution.

Effects of biological interactions

As has been emphasized throughout this book, any attempt at interpreting the complexities of animal and plant communities in soft sediments in terms of physical determinants alone is doomed to failure: interactions between and within species make sure of this. But how do biological interactions affect distribution of the benthic species in estuaries? There is as yet no clear answer, but here we discuss some possibilities.

The influence of food supply

The majority of benthic invertebrates in estuaries are deposit feeders, or at least species that include deposit feeding among a suite of abilities. For example, small snails (e.g. *Hydrobia* spp.), polychaetes (e.g. *Nereis* spp.), and bivalves (e.g. *Macoma*) are common in temperate estuaries. For these animals, food supply may not be limiting because of the enormous supply of detritus to estuaries, both produced nearby and imported from the rivers and from the sea. Bearing in mind, however, the lack of agreement about which fractions of the material broadly labelled detritus are actually important for deposit feeding—bacteria, dead particulate organic matter or diatoms, for instance (p. 80)—food supply may be more variable and limiting than we presently think. For suspension feeding species, the situation is clearer. Suspended sediment may clog filtering mechanisms, as discussed above; and the supply of suspended phytoplankton is very variable, both between and within estuaries. The variation within one small area has been demonstrated in the Eel River estuary, Nova Scotia. Here, as in many other estuaries, much of the plant material in the water column is not true phytoplankton, but consists of benthic diatoms lifted off the mud surface by water currents. Estuarine water passing over areas of rich primary production thus contains much more chlorophyll-*a* than does sea water brought in on the tide, which has only true phytoplankton (Roegner, 1998). Depending upon position in the estuary, benthic bivalves like *Mya arenaria* and *Macoma balthica* receive quite different 'rations' (Fig. 8.8). Do these differences in food supply determine bivalve distribution? As bivalves in the Eel River can actually deplete the water of food at some states of tide, it seems likely that their distribution is partly controlled by food supply—but such hypotheses of course need experimental testing.

The influence of competition

We have already established that competition between species, especially on muddy shores, can be important in determining small-scale distribution patterns (Chapter 4). Because such emphasis has been placed on the effects of salinity as a major controller of distribution within estuaries, there have been few conclusive studies on the large-scale influences of competition. Indeed, it has been argued that competition may be rare in estuarine soft sediments because most populations are held at levels below the carrying capacity of the habitats by other factors (Barnes, 1994b). While this may often be so,

Fig. 8.8 Variations in the concentration of suspended chlorophyll-*a* at three sites in the Eel River estuary, Canada (bottom graph). Note the scale is logarithmic. Chlorophyll remains high at West marsh due to suspension of diatoms normally living on the mud surface. Also shown are salinity and tidal height. (After Roegner, 1998.)

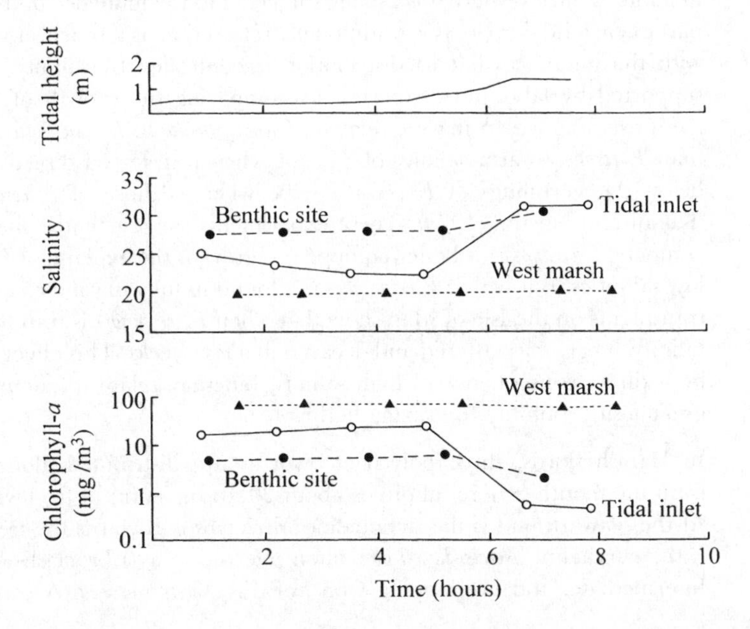

evidence is accruing that competition is sometimes important in estuaries, and three examples are discussed.

Amphipods of the family Gammaridae are some of the commonest invertebrates on marine and estuarine shores. The distributions of species change along the length of estuaries, and have usually been correlated with salinity: the species differ in degrees of tolerance to changing salt concentrations, and some have developed extremely sophisticated mechanisms of osmoregulation (Bulnheim, 1991). While most can take up ions by active transport, some can lower their surface permeability, and some can produce hypo-osmotic (dilute) urine. Are distributions really to do with these differing physiological capacities?

Experiments in the laboratory suggest that interactions between species are important, especially during mating. For example, when the freshwater species *Gammarus pulex* is placed together with *G. duebeni* (which has populations in fresh water and in brackish water), males of both species seize females of the opposite species, keep them in the mating position, and then eat them slowly after the female moults (Dick *et al.*, 1990). By doing so, they prevent males of the 'correct' species from mating with them. The males also guard females of their own species and significantly reduce the incidence of predation. Males of *G. pulex* are more successful guards than those of *G. duebeni*, so in a mixed population, guarding ability may explain why *G. pulex* can displace *G. duebeni*. Observations on mixed populations of *G. duebeni* with another brackish-water species, *G. zaddachi*, show that males of *G. zaddachi* also monopolize females of *G. duebeni* (Remane and Schlieper, 1971). Competitive effects during mating could therefore explain at least some of the sequences of gammarid species along estuaries.

The brown alga *Fucus ceranoides* is not found in fully marine conditions, but inhabits estuaries where it is usually subjected to the influence of fresh water for part of each tidal cycle. It is traditionally referred to as a brackish-water species, with the inference that its distribution is controlled by salinity. This view is supported by laboratory experiments comparing the growth of germlings of *Fucus ceranoides* with a marine relative, *Fucus vesiculosus*: *F. vesiculosus* grows better than *F. ceranoides* at a salinity of 34, but when transferred directly to salinities below 24, germlings of *F. vesiculosus* die whereas those of *F. ceranoides* thrive (Khfaji and Norton, 1979). These experiments suggest that if the two species compete, *F. ceranoides* is better equipped to grow at the expense of *F. vesiculosus* at low salinities. But why is *F. ceranoides* not found in fully marine conditions? Field transplants on the Isle of Man show that when *F. ceranoides* is transferred to high salinity, it becomes tattered and decayed in a few weeks. This effect is unlikely to be a direct consequence of high salinity, but may relate to competition or to greater susceptibility to grazing herbivores.

In Danish fjords, three polychaete worms are distributed along a gradient from the mouth (where salinity is about 20) to the entry of freshwater streams. At the seaward end is the stenohaline *Nereis virens*. Towards the freshwater end is the euryhaline *Nereis diversicolor*, often referred to as a brackish-water species. Intermediate, and overlapping with both, is *Nereis succinea*. A combination of sediment preferences and intolerance of low salinity probably limits penetration of *N. virens* and *N. succinea* landward. But what limits seaward distribution? When the species are placed in experimental tanks, interspecific effects are quickly seen (Kristensen, 1988). Species held individually show low mortalities over a month. But when *N. virens* and *N. diversicolor* are held together, *N. diversicolor* suffers high mortality (Fig. 8.9). These observations suggest that competition from the aggressive species *N. virens* may prevent *N. diversicolor* from moving seawards. There is also competition between *N. virens* and *N. succinea*, which may determine the diffuse boundary between them. Because *N. diversicolor* is the least aggressive of the three species, it is confined to areas of low salinity, where the other two species cannot penetrate.

These three examples show that competition can be extremely important in governing distribution within estuaries—and they suggest extra caution when implying direct effects of salinity on distribution.

The influence of behaviour

Distribution of the benthos in relation to physical variables suggests that species either choose particular physical conditions, or that they suffer greater mortality in 'unfavourable' ones. The study of larval settlement is therefore important, as discussed on p. 123. But many benthic species also move around as adults, and studies of the amphipod *Corophium volutator* show that its distribution is strongly affected by adult movement (Hughes and Gerdol, 1997). Both the Blackwater

Fig. 8.9 Mortality of three species of nereid worms from a Danish fjord, kept in laboratory tanks for 30 days. Histograms show means and bars show the range. When the species were kept separately, mortalities were low. When two species were kept together, one species usually showed increased mortality. (After Kristensen, 1988.)

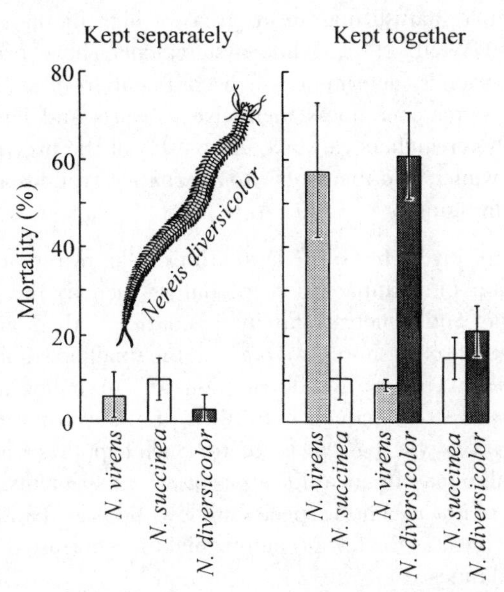

and Crouch estuaries in southeast England have dense populations of *Corophium*. These are most abundant in creeks and semi-enclosed bays, where densities are twice as high as on open flats. Aggregations occur not in relation to sediment type, but in response to dispersion of adults on flood tides which take them into sheltered areas. As populations in different areas swim at different states of tide, local behaviour patterns could result in very different distributions. The study of behaviour patterns in estuarine invertebrates is thus an important topic.

The influence of predators

Estuaries, and estuarine flats in particular, are characterized by an abundance of mobile predators. Wading birds arrive on autumn migration. Fish move in at various seasons, as we shall see in Chapter 9. In addition, there are numerous small infaunal predators and larger epibenthic species. All of these take their toll on the benthos, more so in estuaries than in lagoons (Barnes, 1994b). But this does not mean that predators actually control the distribution of their prey. In spite of the large amounts of prey species eaten, the production rates of the benthos are so high that the effects of predation may be insignificant overall. In particular, vertebrate predators usually attack the largest prey individuals, leaving behind an enormous population of juveniles that quickly grow to take their place. It has therefore been argued that despite the high levels of predation quoted in Chapter 4, the overall effect of bird predation on species such as *Hydrobia* might be less than 0.1% of production (Barnes, 1994b).

Indeed, it seems more likely that control operates the other way round: densities of benthic invertebrates control the distribution of their vertebrate predators. For example, flatfish migrate to areas of high biomass to feed (Wolff *et al.* in Jones and Wolff, 1981), while oystercatchers have preferred mussel beds as feeding grounds, determined by a combination of available prey and density of the oystercatcher flocks themselves (Zwarts and Dent in Jones and Wolff, 1981). Oystercatchers can take up to 40% of the prey population on a mussel bed in a winter, and their move to alternative grounds allows them to maintain efficient foraging.

In terms of invertebrate predation, little is known about effects on the overall distribution of infauna. Some predators such as the crabs like *Carcinus* are euryhaline, and penetrate far into estuaries. Others are stenohaline, like the dogwhelks of rocky shores (*Nucella*) and the small opisthobranch *Retusa* that preys on *Hydrobia*. The relative distributions of euryhaline and stenohaline species may have direct consequences for their prey. For instance, does *Hydrobia* expand its populations where *Retusa* is absent? Such hypotheses await testing. Areas such as the Baltic may be ideal for appropriate experiments. In oligohaline regions, numbers of macrobenthic species are low (Bonsdorff and Blomqvist, 1993), and here the effects of the few predators might be contrasted with predator effects in higher salinities.

Conclusions: factors governing distribution of the benthos

Physical and biological factors

We are not yet in a position to provide any overall view of how to fit together the various factors that determine species distributions in estuaries. In terms of physical influences, however, the use of multivariate analyses is beginning to give much insight. For instance, the macrobenthos of Arcachon Bay, a mesotidal inlet in western France, can be grouped into five assemblages. The distribution of these assemblages relates to a variety of factors, the most significant of which is sediment type, but includes also salinity and water depth (Bachelet *et al.*, 1996). In the Westerschelde estuary, the Netherlands, the distribution of meiobenthos has been investigated in relation to a suite of environmental variables (Soetaert *et al.*, 1994). Here species composition is determined mainly by distance upstream (correlated with salinity, oxygen, and temperature), but also by sediment characteristics (median grain size and silt content), water depth, and depth in the sediment.

It is thus possible at least to see a framework that may be used in the future (Fig. 8.10). Salinity seems to set both upper and lower limits: stenohaline species cannot penetrate very far into low salinities, and euryhaline ones seldom penetrate as far as a salinity of 5. Hence the lowest numbers of species occur

Fig. 8.10 A possible framework to express the effects of the variables that affect distribution of benthic species in estuaries.

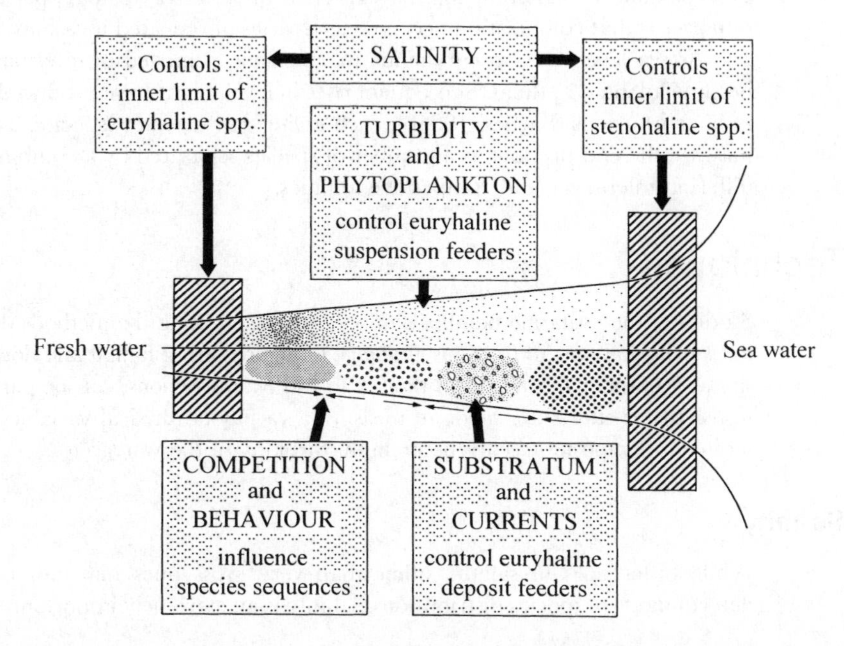

in salinities of 5–8. The overall decline in species numbers in these low salinities probably relates to the general inability of euryhaline marine invertebrates to osmoregulate. Those species that can osmoregulate, such as *Gammarus* spp. , *Nereis diversicolor*, and the oligochaetes such as *Nais* can be spectacularly abundant in low salinities (Little, 1984).

Within the bounds set by salinity, distributions relate to physical variables such as substratum and current regime, oxygen tension, and turbidity. The turbidity gradient also determines phytoplankton production, as we shall see in Chapter 9, and so determines distribution of suspension feeders both directly (via clogging mechanisms) and indirectly (via food supply). Biological interactions such as competition may also be important, particularly in determining sequences of related species, but as yet their effects have not been investigated in detail.

A historical approach

On top of these mechanisms that determine present-day distribution, it is important to examine possible historical explanations for the low diversity in estuaries. Chief among these is the suggestion that because many estuaries are, in their present form, no older than the last ice age, there has been little time for evolution of appropriate species within them. McLusky (1989), for instance, pointed to a comparison between the estuaries of northwest Europe, mostly less than 10 000 years old, and those of Southeast Asia, which are much older—and have a much richer fauna. There is probably some truth in this 'age hypothesis', although it should be noted that tropical faunas are usually richer than temperate ones for other reasons, and the hypothesis is hard to test.

One possible explanation for the existence of series of related species along estuaries is that colonization occurred as a series of repeated invasions from the sea. As sea level rose, so euryhaline marine species moved in to estuaries and became adapted to them. Subsequent rises in level might have produced a series of later invasions (Barnes, 1994b), each giving rise to a new species. Today we may see the end product as a number of species separated by a combination of differing tolerances and competitive abilities.

Techniques

Studies of the estuarine benthos can be undertaken using the methods described for specific habitats in previous chapters. Comparisons of faunal and floral diversity at a number of sites, from the sea to freshwater regions, can be particularly instructive if efforts are made to measure a variety of physical variables such as sediment composition, current strength, turbidity of the water, etc.

Salinity

While differences in salinity other than very gross ones may not be direct determinants of species distributions, salinity is an extremely important 'marker' in estuaries, distinguishing the degrees of marine and freshwater influence. It is also the easiest of the estuarine variables to measure.

In theory, salinity comprises the mass of dissolved inorganic matter in a particular mass of water (Head, 1985). Until 1982, it was expressed as g/kg or ‰. Measuring this gravimetrically has been found difficult, however, and most early measurements were made by measuring the concentration of chloride. Chloride can be measured accurately by titration against silver nitrate, and the figure converted to salinity. With the development of accurate devices to measure conductivity, these have taken over from titrations, and in 1982 a 'Practical Salinity Scale' was adopted, in which conductivity is related to that of standard sea water. This scale is a pure ratio, and thus has no dimensions or units, which is the reason that salinities are now referred to just as a number: e.g. 'the salinity is 35.0'. Salinity in the field is now therefore best measured using a salinometer, which is a conductivity meter graduated on a salinity scale. Note, however, that salinometers may be bulky, and rough measurements of salinity can be made with much smaller devices such as refractometers. Cheapest of all is probably the use of a hydrometer, while silver nitrate titrations are simple but tedious.

9 Life in the estuarine water column

In many ways, physical conditions in the estuarine water column are more rigorous than in the benthos. For example, there is no buffering by the sediment against sudden changes in salinity. But in fact the plankton does not often have to face the problem of great changes in salinity because the organisms travel *with* bodies of water rather than staying still while water of different salinity washes over them. There is also the constant problem that organisms may be washed out to sea. Nevertheless, some estuaries contain flourishing populations of plankton, fish, and swimming crustaceans, and as we shall see they often act as nursery grounds for fish species that mature in the sea.

The distribution of the plankton

Species diversity of both phytoplankton and zooplankton in estuaries, like that of the benthos, decreases with distance from the sea. Nevertheless, successful species may be *very* successful, and these often show well-defined distribution patterns. We begin by discussing these patterns.

Distribution along the length of estuaries

Along any one estuary there is usually a sequence of planktonic assemblages, just as there is for the benthos. At the riverine end are those usually regarded as euryhaline freshwater species. Then come estuarine species, followed downstream by euryhaline marine species, which give way to stenohaline marine species in the fully marine zone. An example is provided by phytoplankton in the Navesink estuary, eastern USA, where freshwater diatoms and cyanobacteria occur up to a salinity of 5; estuarine diatoms, euglenoids, and dinoflagellates provide variable assemblages between salinities 10 and 20; euryhaline diatoms and dinoflagellates are found up to a salinity of 30, and stenohaline species above this. Similar kinds of sequence can be found in the phytoplankton and zooplankton of estuaries worldwide, although the picture is often more complicated. For example, in the Brisbane River estuary, Australia, copepods of the genus *Gladioferens* are abundant. *G. spinosus* lives at the freshwater end of the estuary but is replaced by the very euryhaline *G. pectinatus* over most of the length of the estuary. At the seaward end, *G. pectinatus* is replaced by stenohaline marine copepods, but the boundary between these and *Gladioferens* varies with differing combinations of temperature and salinity over the seasons (Green, 1968).

In the Severn estuary, four distinct assemblages of zooplankton species occur along the length of the estuary (Collins and Williams, 1981). At the upstream, low-salinity end are estuarine species such as the copepod *Eurytemora affinis* accompanied by the ctenophore *Pleurobrachia.* These occur through the length of the tidal river up to a salinity of 30. An 'estuarine-marine' group, typified by the copepod *Acartia bifilosa*, together with mysids, occurs seawards of *Eurytemora*, up to a salinity of 33.5. Euryhaline marine species such as the copepod *Centropages hamatus*, with the polychaete *Tomopteris*, occupy salinities between 32 and 34. Finally, the copepod *Calanus helgolandicus* with other copepods and the chaetognath *Sagitta elegans* form a stenohaline marine group that does not usually penetrate into the estuary from the sea beyond the 33 isohaline (Fig. 9.1).

In brackish seas such as the Baltic, species distributions can be even more strictly related to salinity, and euryhaline marine species penetrate further into dilute regions. The central Baltic, which has surface salinities of less than 10, contains medusae such as *Sarsia*, the ctenophore *Pleurobrachia*, the chaetognath *Sagitta* and many other marine-derived species. Here many planktonic groups such as the rotifers, copepods, ostracods, and cladocerans can be classified as typical of relatively restricted salinity zones (Remane in Remane and Schlieper, 1971).

Fig. 9.1 The distribution of four copepods in the Severn estuary, showing the areas in which each species was found at more than 10 individuals/m^3 in August 1974. The four species, *Calanus helgolandicus, Centropages hamatus, Acartia bifilosa,* and *Eurytemora affinis,* look very similar but have different, though overlapping distributions. (After Collins and Williams, 1981.)

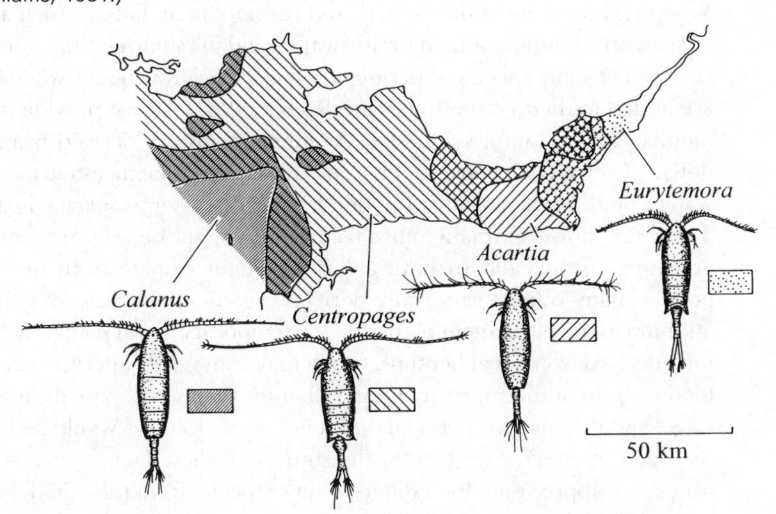

Explanations of plankton distribution

Classical explanations of why plankton are distributed as described above rely, naturally enough, on the supposition that salinity itself is, in some way, the causal factor, and this explanation is still accepted by many. It is supported by a great weight of physiological evidence which shows that most species have distinct salinity ranges over which they will survive, and 'salinity optima' at which they grow, survive, reproduce, or feed best. There is no doubt that major changes in salinity affect planktonic flora and fauna: the majority of marine plankton cannot survive when salinity drops below 30, just as few freshwater plankton can survive when it rises above 5. At this gross level, salinity provides an apparently reasonable explanation for plankton distribution.

When it comes to pinning down details of distribution, the picture is much more unsatisfactory. Species in nature seldom extend to the ends of their physiological salinity ranges, as we have pointed out when discussing the benthos, suggesting that other factors intervene. Such restriction of the 'potential niche' to a 'realized niche' has been a major tenet of ecological theory for a considerable time. Some species occupy different salinity ranges in different regions, making the overall picture more complicated. Emphasis should therefore now be placed on considering the other variables that determine distribution.

In the marine environment, one of the major phenomena that has not so far been explained is the observation that 'neritic' or coastal waters contain different planktonic species from those of the open ocean. The classical boundary between neritic and ocean waters in the Atlantic, for instance, is that between two chaetognaths, *Sagitta elegans*—found in neritic waters—and *Sagitta setosa*—found only in the open ocean. Open-ocean water has less nutrients than neritic water, but this is unlikely to affect predators like chaetognaths, at least not directly. There is no major difference in salinity, so this cannot provide an explanation, although many neritic species do penetrate into estuaries. We have seen that *Sagitta elegans* occurs in the Severn, and cosmopolitan diatoms such as *Skeletonema costatum* are common both in neritic waters and in estuaries. One intriguing point here is that some species of neritic dinoflagellates grow better when humic acids are added to their culture medium (Round, 1981)—these provide materials the dinoflagellates cannot synthesize. As humic acids are derived from the breakdown of terrestrial detritus, they are most common in estuarine and neritic waters, and might account for better growth of some species in these areas. There are, however, many other factors that might be relevant—neritic waters are more turbid, and so have a different light climate from the ocean; they possess many other terrestrially derived organic molecules; and they have, in addition to the permanent plankton, the temporary larval plankton ('meroplankton') derived from local benthos, which may completely alter the structure of the food web. In addition, many neritic diatoms themselves pass through a benthic stage, and this may be a crucial difference, as such stages would be impossible in the open (deeper) sea. Overall, the sum of all these factors means that neritic waters are appropriate for a different set of species from those in the open ocean.

One example from the zooplankton shows that placing too much emphasis on the influence in estuaries of salinity *per se* may be misleading. The copepod *Acartia bifilosa* is a euryhaline marine species typical of estuaries in Europe, and replaced by *Acartia clausi* in more marine waters. But on the east coast of the USA, *A. clausi* is considered an estuarine species typical of intermediate salinities. While it is therefore possible that salinity is the over-riding factor affecting *Acartia* distribution, it seems that, as with the benthos, there are too many confounding variables to be sure of this.

In the Severn estuary, as described above, the zooplankton, and particularly the copepods, can be classified into assemblages based on different salinities. Yet as salinity increases, turbidity falls, phytoplankton species diversity increases, nutrient constituents change in concentration and proportion, and the influence of substances originating from river run-off declines. These will all influence one of the major factors for the zooplankton—the food supply. Most marine copepods are herbivores or carnivores, but the two species found furthest upstream, *Acartia bifilosa* and *Eurytemora affinis*, are omnivores and may well depend more upon detritus for food. Indeed, there is so little phytoplankton in the upper reaches of the Severn—often less than 1 mg chlorophyll-*a*/m^3—that true herbivores would find life impossible. On the other hand, bacterial levels can be very high—as many as 2×10^5/ml. Such bacterial abundance, associated with suspended sediment, may therefore account for the fact that detritivores replace herbivores in the upper reaches. In other localities, where chlorophyll-*a* production is strongly seasonal, plankton communities change in composition in relation to food supply. For example, in the Elbe estuary, Germany, detritus-eating rotifers dominate early in the year, but are replaced by phytoplankton eaters later on (Holst *et al.*, 1998).

In summary, estuarine and lagoonal assemblages of both phytoplankton and zooplankton differ widely from marine and freshwater ones. The factors that control their composition are, however, badly known and certainly involve more than simple differences in salinity tolerance. In particular, little is known of the ways in which biological interactions such as competition and predation may reduce the potential niches of plankton that can tolerate a wide range of physical conditions.

What controls phytoplankton growth?

Estuarine waters are remarkably variable in their productivity. For example, primary production by phytoplankton in the Zaire estuary seldom exceeds about 30 gC/m^2 per year, while in the Danube estuary rates may be 50 times as high. To understand why this should be so, and hence to understand why some estuaries have abundant life in the water column, often with extensive fish stocks, while others appear depleted, we need to consider the various factors that influence phytoplankton growth.

Phytoplankton grow by fixing inorganic carbon and turning it into organic compounds. To do this they require access to a carbon source (usually dissolved carbon dioxide or bicarbonate), enough light to drive the process of photosynthesis, and a supply of various chemical nutrients, particularly nitrogen, phosphorus, and silicon. There is seldom if ever a shortage of available carbon, so the supplies of light and nutrients usually determine how fast phytoplankton grow and replicate. In classical studies the influences of these factors have been described in the sea, and we start there before considering how different the situation can be in estuaries.

Phytoplankton growth in the sea

Light intensity declines exponentially from the surface of the sea downwards, because light is absorbed by the water and by dissolved organic compounds, and scattered by suspended particles. Blue–green light penetrates furthest, but even in clear ocean water only 10% of the light penetrates to about 100 m. This may be sufficient to allow some photosynthesis, but usually the phytoplankton grow in a shallower layer than this. We can see why if we consider how conditions in the ocean change over the seasons. During winter and spring, the top 100 m is freely mixed at temperate latitudes. The mixing ensures constant renewal of nutrients, but even so, not much photosynthesis occurs in winter because of the low light regime—and because mixing processes also carry some of the phytoplankton out of the photic zone. When spring arrives, and overall light levels rise, the phytoplankton take advantage of the nutrient supply and reproduce rapidly, often causing a 'spring bloom'. As summer progresses, the grazing zooplankton begin to make an impact on phytoplankton populations, and as the surface waters heat up a vertical temperature gradient develops. Usually the top 15–40 m in temperate oceans remains relatively uniform in temperature, but below this level there is a zone of rapid temperature decrease—the thermocline. While there may be only a few degrees difference in temperature across the thermocline, the difference causes a small density gradient. This temperature-induced stratification prevents further vertical mixing, and under these circumstances the phytoplankton may use up the supply of nutrients in the surface water and become 'nutrient-limited', until the thermocline breaks down in the autumn gales. There may then be another small burst of phytoplankton reproduction before falling light levels bring a halt to the year's growth.

This over-simplified account ignores the importance of mixing at 'fronts', and also the important point that in fact phytoplankton can often continue to grow even when nutrient concentrations are very low. Control of growth may have much to do with the small-scale distribution of nutrients, the storage of nutrients by phytoplankton, and with micro-environments associated with suspended particles (Grahame, 1987). As in the benthos, habitats in the plankton have a fractal distribution, and small species can take advantage of conditions not perceived by larger ones. This may be particularly important because a large fraction of production is determined not by 'classic' species such as diatoms, but by very small organisms—the 'nanoplankton' (species 2–20 μm in size) and the 'picoplankton' (those between 0.2 and 2 μm). Nevertheless, the cycle of phytoplankton growth in temperate oceans is controlled mainly by the degree of mixing of surface waters with deep, nutrient-rich waters, and the major cause of stratification is a temperature gradient.

Phytoplankton growth in estuaries

The situations in estuaries and lagoons are different from those in the sea for several reasons, and vary widely between different hydrographic regimes. At one extreme, in macrotidal estuaries where river water and sea water are well mixed, primary production tends to be low. At the other, in microtidal estuaries where mixing is prevented by stratification, production can be extremely high (Monbet, 1992). We consider a series of situations, in which the interactions between mixing, nutrient supply, turbidity, and rate of flow through the system determine the timing and degree of production.

In the Zaire estuary, west Africa, the freshwater plume is well mixed with sea water by the tides, so that nutrient concentrations are high throughout the water column, and there is no stratification. Because of the high turbidity associated with violent tidal mixing, however, light levels in the estuary are low and there is little primary production. Most production occurs at the outer edge of the estuary (Fig. 9.2), where turbidity has fallen and salinity is high; but by this time the nutrient-rich plume has been diluted with sea water and the overall rate of production is low because of nutrient limitation. Primary production in the Zaire estuary is only 3.7–37 gC/m^2 per year (Humborg, 1997).

Fig. 9.2 The distribution of primary production in well-mixed and stratified estuaries. In well-mixed estuaries, high turbidity keeps production low even if there are high nutrient levels. In stratified estuaries, the high turbidity of the rivers clears over the salt wedge, allowing production, but this falls at the seaward end of the plume as nutrients are used up. (After Humborg, 1997.)

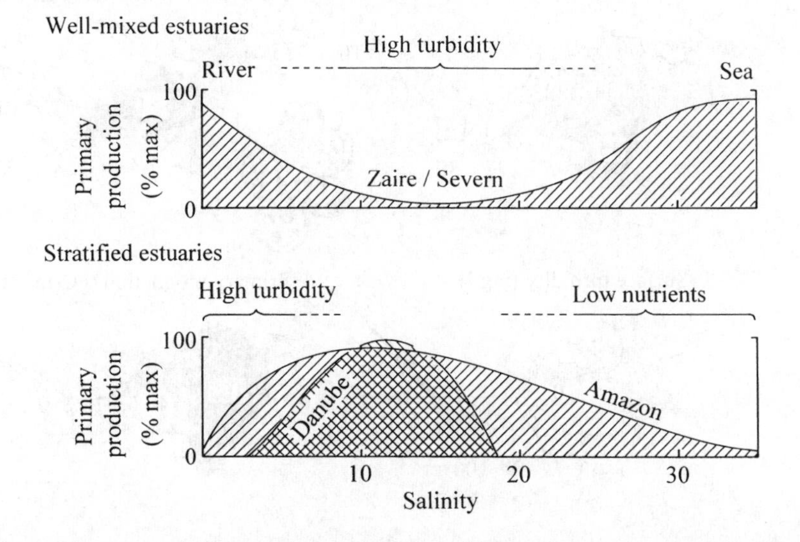

Conditions in the Severn estuary/Bristol Channel complex in the UK are similar, but allow us to add more detail (Fig. 9.3). Here the high turbidity again prevents light penetration in upstream regions, and although violent mixing brings phytoplankton near the surface regularly, the algae never stay there long enough to photosynthesize at more than a maintenance rate. Overall production is therefore low, but increases downstream as conditions for phytoplankton improve. There are no spring blooms in the Severn, although surprisingly there may be summer blooms (Fig. 9.3). These occur because some algae, such as *Phaeocystis*, can adapt to low light intensities, and then multiply. The blooms can therefore become dense, although they do not last more than a month (Joint, 1984).

The same limitations by low light levels upstream and low nutrient levels downstream can occur in rather different situations in the tropics. In Queensland, Australia, the Logan River runs into shallow Moreton Bay, where fresh water and sea water are well mixed. Here the experimental addition of nitrate to bay waters increases productivity from its normal level of about 200 gC/m^2 per year, showing that the phytoplankton are normally nutrient limited (O'Donohue and Dennison, 1997).

This relatively simple picture, present when waters are well mixed, is complicated by the presence of stratification. As we have seen in Chapter 7, the degree of mixing in estuaries usually depends upon the relative strengths of river flow and tidal currents. The resultant stratification, if any, is usually determined not by a temperature gradient (the thermocline) but by a gradient

Fig. 9.3 Distribution of salinity, turbidity, and phytoplankton in the Severn estuary. Conditions are given for summer, but surveys were not undertaken at the same time (salinity in August 1974; turbidity in August 1980; and chlorophyll-*a* in June 1974). (After Joint, 1984 and Collins and Williams, 1981.)

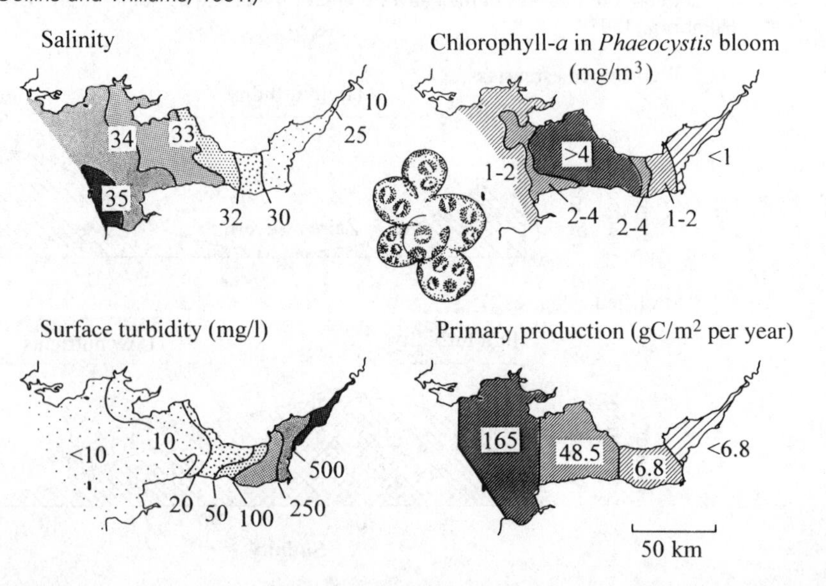

in salinity, known as a halocline. Haloclines can be both more stable and more abrupt than thermoclines because a small change in salinity produces a much larger change in density than does a small change in temperature. For example, the temperature change across a thermocline might be 5°C, giving a difference in specific gravity of about 0.001. But salinity change across a halocline may be 10 or more, giving a difference in specific gravity of about 0.008. Across a salt wedge, the density difference between sea water and fresh water is about 0.02. The force needed to break up a typical halocline is thus much greater than that needed to break up a typical thermocline. This means that haloclines can be stable even in shallow waters, while shallow thermoclines are rare because they are broken up even by slight water movement.

In the Chesapeake Bay, a partially mixed estuarine system, the central reaches have surface salinities of 10–17, while bottom waters are 20–21. The steepest part of the halocline varies from 5 to 15 m down, and a spring phytoplankton bloom occurs just above it, rather than near the surface. This is possible because the water is clear, unlike the turbid waters of fully mixed estuaries. Phytoplankton biomass accumulates readily in the spring, because the grazing zooplankton are still low in numbers. Indeed, so much biomass settles to the bottom that when it decays it leads to low oxygen levels below the halocline: the Chesapeake is eutrophic. Later in the summer, primary production reaches a peak in surface waters, but biomass actually declines—populations of grazers have built up by now, and they remove the phytoplankton faster than it can replicate. There is also some evidence that phytoplankton are nutrient limited at this time—lack of nitrate, phosphate, and silicate can all in turn limit production when they are used up by algal cells and lost below the halocline. Nevertheless, production levels reach 400 gC/m^2 per year—much greater than in the sea, or in well-mixed estuaries (Malone *et al.*, 1996).

Where stratification is more extreme, the high levels of nutrients in river water can lead to very high rates of production. In the Danube estuary, which runs into the Black Sea, the shallow surface layer, derived from the river, has high concentrations of nitrate and phosphate, and supports a high rate of primary production. This is possible because as the fresh water floats on the salt wedge, sediment precipitates down through the wedge and allows light to penetrate freely, sometimes as much as 18 m (Humborg, 1997). In the Danube, maximum production thus occurs at intermediate salinities, compared with a low rate in fresh water—which is too turbid—and a low rate in the sea—which is nutrient limited (Fig. 9.2). The overall levels of production are extremely high in this situation—up to 1600 gC/m^2 per year.

Extreme stratification can also develop in sheltered lagoons. Swanpool, for instance, has a well-marked halocline throughout much of the year at about 1 m depth. In the spring, most phytoplankton production occurs in this halocline, not at the surface—a situation comparable with the Chesapeake

Fig. 9.4 Stratification in the Swanpool, a coastal lagoon in Cornwall, UK, showing associated distributions of phytoplankton. The dinoflagellate *Gyrodinium resplendens* gathers in the halocline in spring, but is more widely distributed in summer. The diatom *Chaetoceros muelleri* has a more even distribution with depth throughout the year. Phytoplankton abundance kites show distributions as percentages of the maximum density. Note that maximum biomass is in fact more than twice as high in summer than in the spring. (After Crawford *et al.*, 1979.)

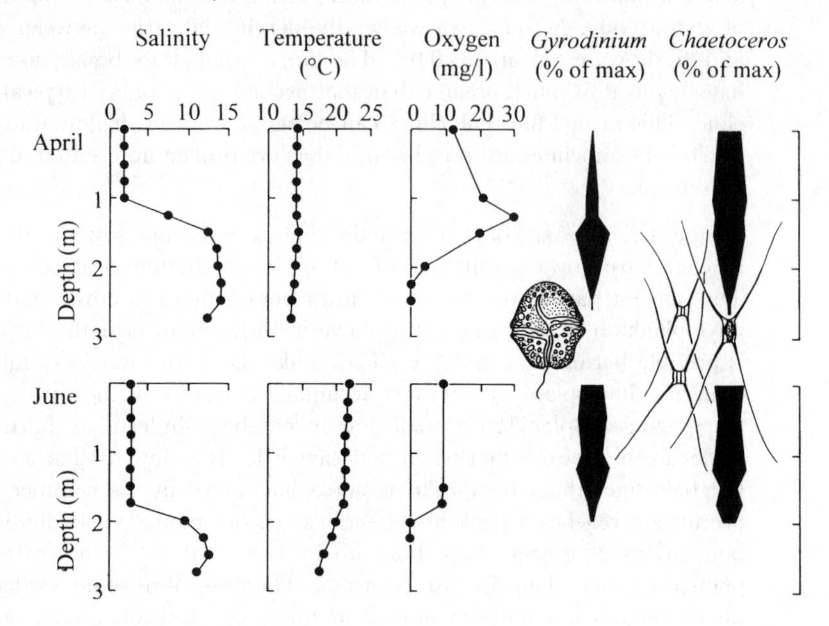

Bay. Dinoflagellates and some diatom species gather here, and their active photosynthesis raises oxygen levels to three times that of air saturation (Fig. 9.4). The Swanpool situation differs from the Chesapeake, however, in that the spring bloom is only small. Biomass increases only later in the year: there are few grazing zooplankton in Swanpool, because of the isolated nature of the lagoon. The late-summer bloom that occurs occupies all the water above the halocline (Fig. 9.4). At this time, nutrients such as nitrate and phosphate are much depleted, and there may be some slight growth limitation due to lack of phosphate (Crawford *et al.*, 1979). In general, however, the small size of the lagoon means that there is sufficient renewal of nutrients in the inflowing water to provide for phytoplankton growth. A similar situation occurs in Perdido Bay, Florida—a microtidal estuary with much nutrient enrichment from land run-off. Here the phytoplankton are only occasionally nutrient limited, as shown by the experimental addition of nitrate and phosphate (Flemer *et al.*, 1998). In general, nutrient limitation is thus relatively rare in estuaries and lagoons because, unlike the condition in the sea, nutrients are continually being replaced.

The zooplankton and their predators

Keeping pace: how not to get washed out to sea

By definition, planktonic organisms have insufficient swimming ability to allow them to counteract water currents, so it may seem surprising that they can exist in estuaries, where the overall flow runs out to the sea. However, estuarine water rarely flows directly seawards, as the descriptions in Chapter 7 have implied. Instead, it circulates in large rotatory 'gyres' or mixes in turbulent eddies, and in any case estuarine water often has a long flushing time—tidal movement to and fro may be great, but the time it takes any one particle of water to travel from river to sea is frequently measured in weeks.

It is true that zooplankton densities are often low, and this has been attributed to the removal of populations in a seaward flow happening faster than the organisms can reproduce. While phytoplankton have rapid replication times, so that populations can double in a day, zooplankton usually pass through a series of larval stages and may take weeks or months to complete the cycle from adult to adult. Consequently, the problem of being washed out of estuaries is greater for them. Nevertheless, estuaries with clear water and with high phytoplankton growth often support large zooplankton populations. How do they manage this? A combination of active movement and passive distribution may account for their retention. First, most zooplankton undergo vertical migrations, sinking down during the day and rising at night. These active movements may allow them to exploit inflowing and outflowing currents at various depths. For example, the copepod *Eurytemora affinis* swims up into the water column during flood tides, but sinks to the bottom on the ebb. This is not a day–night alternation, but one geared to the tides, and it allows the copepod to maintain populations in the Conwy estuary, Wales (Hough and Naylor, 1991). A second mechanism, the passive movement upstream of particular size fractions, has been demonstrated in many estuaries. Thus the residual upstream flow in salt wedges results in accumulation of copepods at the upstream limit of the saline wedge in the Gironde estuary (France) and in the Chesapeake Bay (Heip *et al.*, 1995). In the Newport River, North Carolina, passive upstream drift helps to retain larvae of the mud-crab *Rhithropanopeus* within the estuary (Chen *et al.*, 1997).

Utilizing resources: the planktivores

Where there are large populations of zooplankton, planktivorous fish may be abundant. These species usually enter estuaries seasonally, take advantage of the rich food source, and migrate out again to breed at the estuary mouth or in the sea. A good example is provided in the Apalachicola estuary, Florida, which is a river-dominated, bar-built estuary (Sheridan and Livingston in Livingston, 1979). Here there are abundant populations of calanoid copepods and mysids, and these are exploited predominantly by the bay anchovy, *Anchoa mitchilli.* This species spawns within the estuary, and so is present all through the year, but is most abundant in autumn. As the anchovies grow in size, they change diet to take more juvenile fish, so can be thought of as rising in trophic level (Fig. 9.5).

What factors affect the populations of such planktivores as the bay anchovy? First, they might be expected to compete with other fish species. However, there is some evidence that feeding strategies have evolved to reduce such competition. In North Inlet, South Carolina, the bay anchovy and three other fish species all consume copepods, mysids, and crab larvae, but they take different proportions of these. And when consumption is examined in detail, different fish species are seen to utilize different levels in the water column, and to take different species of copepods (Allen *et al.*, 1995). Second, the planktivores are themselves the prey of larger fish. The bay anchovy is eaten by the sand seatrout, *Cynoscion arenarius*, which invades the estuary in summer, and at this time the anchovy populations decline. Trophic interactions higher up the food web, as well as provision of food supply, may thus be very important in estuarine fish populations.

Brackish seas, such as the Baltic, may contain large populations of planktivorous fish. Here the herring, *Clupea harengus*, supports a large fishery with a catch of 400 000 tonnes annually. Do such large numbers of predators have significant effects on zooplankton populations? Herring eat a variety of planktonic species, especially copepods, but in the Baltic they also consume mysids and benthic amphipods. They are responsible for consuming up to 8% of zooplankton production, and they may take more of mysid production. As mysids also feed on the benthos, the herring have an indirect effect on benthic production as well as on the plankton. The total effect of herring on the benthos has been

Fig. 9.5 Changes in the feeding pattern of the bay anchovy, *Anchoa mitchilli*, with age, in the Apalachicola Bay estuary, Florida. (After Sheridan and Livingston in Livingston, 1979.)

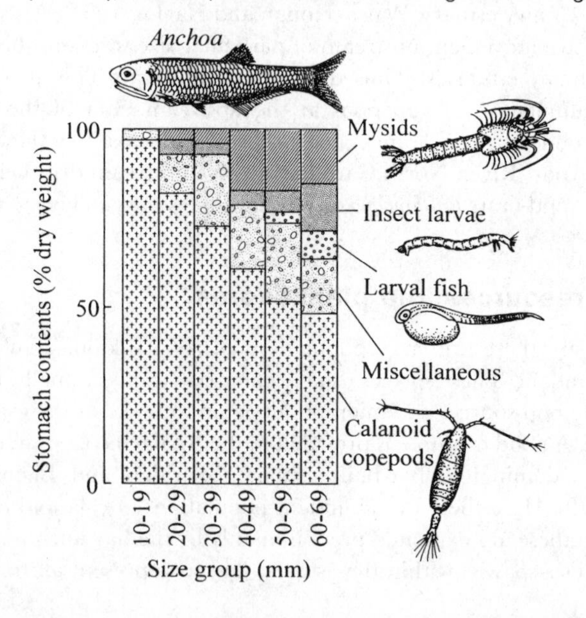

estimated to lie between 20 and 80%, depending upon a variety of assumptions (Aneer, 1980).

In tropical estuaries, there may be a high diversity of larval fish which themselves form a significant part of the zooplankton, the 'ichthyoplankton'. In the estuaries of Sarawak and Sabah (Malaysia), for example, there are many species of clupeiforms (herring family) and gobies. But the generalization that greater stratification leads to more primary production and that this fuels more zooplankton and hence more planktivorous fish does not hold here. In Sarawak and Sabah, turbid macrotidal estuaries have *higher* zooplankton levels and *more* fish larvae than the clearer microtidal estuaries which are stratified (Blaber *et al.*, 1997). Clearly, there is still much to be learnt about the factors governing estuarine planktonic systems.

Estuaries as nurseries for fish and crustaceans

Types of fish in estuaries

The majority of fish found within estuaries are not, like the bay anchovy and herring discussed above, planktivores. Instead, they feed primarily on the benthos, and thus live a 'demersal' existence, i.e. just above the bottom. These estuarine fish can be divided into a number of groups. The 'estuarine-dependent' (or alternatively estuarine-opportunist) species typically enter estuaries or shallow inshore waters from the sea for a period each year, but do not stay there permanently. These are usually the most abundant fish. In addition there are 'marine stragglers' which enter estuaries only irregularly, and are often restricted to the seaward end. While these greatly increase species diversity, they are usually low in numbers of individuals. Coming from the riverine end are the freshwater species, such as rudd and roach, but these are mainly found in low-salinity waters. The truly estuarine fish—estuarine residents—also comprise only a small number of species, although they may provide a high biomass. Views on the composition of this group vary, as many 'residents' leave the estuary for deeper water during winter, and some may breed outside the estuary. Perhaps the gobies are most typical, as they are found throughout the year in estuaries worldwide. Finally, there are migrant species which use the estuary as a route from rivers to the sea or vice versa, and we discuss these later (p. 185).

The majority of estuarine opportunists drift into estuaries as larvae from eggs spawned in coastal waters and when as young fish they become demersal, they take advantage of the rich benthic food sources we have seen in sublittoral sediments, on intertidal mudflats and in salt marshes and mangroves. Estuaries and coastal waters worldwide thus contain immense numbers of '0 group' (i.e. less than 1 year old) fish that use them as nursery grounds before emigrating to the open ocean as recruits to their adult populations.

In the Forth, for example, a macrotidal estuary in Scotland, 36 fish species have been recorded, most of these being either marine opportunists or marine stragglers. Juvenile flatfish are common: dabs move into the estuary for only a few months while plaice stay for nearly a year, and flounders are classed as resident (Elliott *et al.*, 1990). Juvenile gadoids are also common: whiting come in at about 50 mm long in May and emigrate a year later at up to 180 mm, and cod have a similar cycle.

In the Severn estuary, estuarine opportunists make up more than 90% of the fish catch (Potter *et al.*, 1997). These species, such as whiting, flounder, bass, and mullet, are all euryhaline, but are often associated with times or areas where salinity is relatively high. Each peaks in abundance at a specific time of year, when the numbers of 0 group fish dominate. These fish are mostly from 40 to 100 mm in length, and their presence reflects the period of spawning by the adults outside the estuary a few months earlier (Claridge *et al.*, 1986).

Because different species migrate in and out of estuaries over the seasons, the proportions of various types can alter drastically. For example, in Alligator Creek, a tropical estuary in northern Australia, estuarine opportunists dominate in the wet season, but in the dry season these species move out and estuarine residents make up 90% of fish numbers (Robertson and Duke, 1990). Seasonal changes can have other drastic effects, as in temperate estuaries of Australia and southern Africa (Potter *et al.*, 1990). Here some of the estuaries become temporarily closed by sand bars at their mouth, and are effectively lagoons. Many marine species continue to live there for some time, but if closure persists, populations may suffer either because of massive dilution by river water, or because of increased eutrophication. Thus for both estuaries and lagoons, only those with a regular opening to the sea provide effective nursery areas. We should note, however, that while many tropical fish move into estuaries as juveniles, many others move to shallow inshore marine areas which act as equivalent nursery grounds (Blaber, 1997).

Distribution of fish in estuaries

What governs the distribution of fish species in estuaries? As with the benthic and planktonic communities that we discussed earlier, this is a complex question to answer, but at least three factors should be considered. First, fish respond to the physical and chemical characteristics of their habitat. Second, they require a food source. And third, they may need to find refuges from predators. We can now discuss these interacting influences.

Estuarine fish are often categorized by their salinity tolerance, but as we have emphasized before, concentration upon one physical variable is usually misleading. An analysis of flatfish distributions in a Louisiana estuary, Barataria Bay, shows well how many factors affect their choice of microhabitat (Fig. 9.6). Here seven species of flatfish have estuarine habitats characterized mainly by depth, salinity, and seasonal temperature. For example, offshore tonguefish are found in deep, cool saline water, while bay whiff concentrate in warmer, fresher shallow waters. In this example, however, each 'factor' labelled on the graph

Fig. 9.6 Use of microhabitats by flatfish in Barataria Bay, Louisiana, as determined by a factor analysis that resolved six variables into three factors. Each factor is strongly associated with the variables labelled on the axes, but also includes influences from others. The location of each species is plotted as the centre of a balloon, and the diameter of each balloon represents one standard error of the mean. (After Allen and Baltz, 1997.)

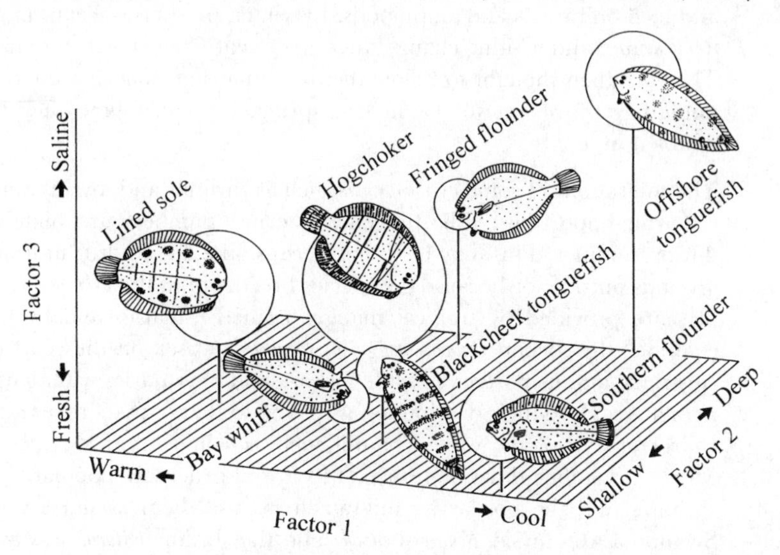

actually conceals other variables—factor 3 represents both salinity and substratum, and so on. A similar study using multivariate analyses in the Humber estuary, UK, concludes that here salinity is the most important variable, closely followed by temperature. Oxygen levels are also relevant, but perhaps surprisingly turbidity has little effect on bottom populations (Marshall and Elliott, 1998). The definition of such physical characteristics of niches can evidently tell us much about major influences on fish distribution. It is essential to note, however, that in the Humber study the physical factors explain less than 20% of the variability in fish numbers—there are presumably many other relevant variables, some at least 'biological'.

Food supply is perhaps the most obvious of these biological factors. Among the benthic feeders are the mullet, which graze algal films from soft substrata. In tropical African estuaries, for instance, each mullet species has a preferred range of sediment types that it grazes, with particle sizes varying from 0.1 to 1.0 mm. The species avoid competition by selecting these different sand types (Blaber, 1997). Some of the estuarine opportunists feed exclusively on benthic infauna and so their distributions depend upon the distribution of the benthos. For example, the flounder, *Platichthys flesus*, consumes the amphipod *Corophium*, the snail *Hydrobia*, and the bivalve *Macoma*, and is most common where these dominate the benthos. But most eat mobile epifauna as well, and include in their diet semi-planktonic species such as mysids. The bass, *Dicentrarchus labrax*, specializes on the shrimp *Crangon* and mysids, while the whiting, *Merlangius merlangus*, has a broad diet including *Crangon*, mysids, amphipods, and cumaceans

(Henderson *et al.*, 1992). The importance of an abundant food supply for the invasions of estuaries by adult estuarine opportunists is nicely exemplified by movements of the whiting in the Severn (Henderson and Holmes, 1989). 0 group individuals, only some 70 mm long, move into the estuary in summer, and feed on mysids and amphipods. In winter, most crustaceans migrate out to deep water, and whiting change their diet to eat *Crangon*, which remains behind. They are then thought to follow the movements of *Crangon* around the estuary until they move out to sea in late spring, by which time they have nearly doubled in length.

The migrations of some crustaceans such as shrimps and mysids mimic those of estuarine-opportunist fish. In general terms, numbers are high in estuaries during summer. This may be either because the larvae drift in from spawning grounds outside, or because the species breed within the estuary. Classic examples are provided by tropical mangrove-lined estuaries and by lagoons. For example, the shrimp *Penaeus merguiensis* spawns at sea off the coast of northern Australia, and its postlarvae settle inshore and in estuaries with dense fringes of mangrove. Here their distribution is related to that of mangroves rather than to characteristics of the water column such as salinity (Vance *et al.*, 1990). The Mexican lagoons described on p. 148 are also densely populated by penaeid shrimp—and the postlarvae migrate in to use them as nursery grounds. In Swanpool, the mysid *Neomysis integer* and the shrimp *Palaemonetes varians* winter offshore but then enter the pool in spring and reach high densities during the summer months (Barnes *et al.* in Jefferies and Davy, 1979).

A survey extending over 1 year recorded 55 fish species and six crustaceans in the Schelde estuary, Belgium. Waves of juveniles move into the estuary to use the area as a nursery ground. Besides finding a rich food source, do they also find a refuge from predators? There are some suggestions that high turbidity may reduce the effectiveness of predators, so the crustaceans may use turbid zones such as the Schelde to escape from predation by species such as the whiting (Maes *et al.*, 1998). Where turbidity is low, crustaceans seek refuge from their fish predators in very shallow water, as the larger predators tend to remain in deep water. In the Rhode river, a subestuary of the Chesapeake Bay, shrimps like *Palaemonetes* and *Crangon* are most abundant in water shallower than 70 cm, where fish and the larger crabs seldom penetrate (Ruiz *et al.*, 1993). In the tropics, it is probable that shrimps—and fish as well—use the structural complexity of mangrove roots and pneumatophores as protection from predators, although experimental proof of this is still lacking (Vance *et al.*, 1990).

Large, turbid macrotidal estuaries in the temperate zone experience influxes of adult fish as well as the drift of larvae. Are these influxes part of large-scale movements in the ocean? Studies in the Bay of Fundy suggest this may be so. When American shad, *Alosa sapidissima*, were tagged in the Bay, their subsequent recapture points spanned the whole length of the North American coast from Florida to Labrador (Dadswell and Rulifson, 1994). The study of fish in estuaries must therefore be intimately bound up with study of nearby oceans.

The migrants

Many estuaries experience seasonal migrations of fish that do not stay for long periods like the estuarine opportunists, but use the estuary as a route between rivers and the sea. Most of these species are 'anadromous', meaning that they spawn in fresh water, and the young hatch and spend their juvenile life there before migrating back to the sea. Typical examples are the lampreys (*Petromyzon* and *Lampetra*), the shads (*Alosa*), the salmon (*Salmo* and *Onchorhynchus*), and the sturgeons (*Acipenser*). Many of these migratory species undergo changes of body colour and spectacular changes of physiology to allow them to cope with the change of external medium (Maitland and Campbell, 1992).

Most of these anadromous fish probably have little influence in estuaries because they spend little time there, but many are highly prized by anglers and fishermen and have substantial monetary value. Atlantic salmon, *Salmo salar*, which reach maturity in the seas near Greenland, migrate back to European and North American rivers after 2 or 3 years. As they can reach over 10 kg, they are prized as sportfish, but they are also trapped, in some cases in basket-work fisheries that date back to medieval times. Shads form a substantial seasonal fishery in the Chesapeake Bay, as they migrate upstream to breed in spring. Sturgeons have been valued commercially for centuries because their eggs are the source of caviar.

Fish that perform the reverse migration, and breed in the sea, are called catadromous. The most well-known are the eels, *Anguilla*. Both the American eel, *A. rostrata*, and the European eel, *A. anguilla*, breed in the Atlantic, east of the Caribbean, but in slightly different areas. Larvae of *A. rostrata* drift northwest to the eastern coasts of North America, while those of *A. anguilla* drift east to Europe, each taking about a year to reach their destination (Maitland and Campbell, 1992). They arrive as juveniles called elvers, and migrate upstream in dense columns. They may stay in fresh water for any time up to 20 years. Breeding migrations then take the adults back to sea, and across the Atlantic to the areas east of the Caribbean, where they die after spawning. From the point of view of estuarine ecosystems, eels may be more important than most anadromous fish, because many individuals stay as estuarine residents, where they form an integral part of the food web.

Techniques

Traditional ways of sampling estuarine plankton and fishes involve the use of boats, and as estuaries are often subject to dangerous currents and tidal conditions, expert boat-handlers are needed. Larger phytoplankton can then be sampled with nets (mesh size 30–60 μm), but for the smaller forms a water-sampling bottle is essential (Tett in Baker and Wolff, 1987). Most zooplankton can be caught by net (mesh size 60–300 μm), but for larger crustaceans and for fish, it is necessary to use seines or trawls (Potts and Reay in Baker and Wolff,

1987). One complication here is that sampling the water column takes a finite time, and especially in large estuaries the water may move considerable distances during a sampling period. Samples should therefore be referred to positions as they would be at a particular state of the tide.

Recently, advantage has been taken of the siting of power stations on the fringes of estuaries. Their intakes of cooling water provide a continuous sampling mechanism, so they can provide much more long-term information than any individual survey (Claridge *et al.*, 1986).

10 Estuarine ecosystems

In previous chapters we have emphasized the variety of habitats within estuaries. These habitats do not, of course, exist in isolation, but interact with each other, both physically and biologically. For example, mangroves and salt marshes provide organic detritus that may fuel food webs on adjacent flats, while fish and crustaceans move backwards and forwards between the mudflats and the marshes. Indeed, fish may spend much of their time outside estuarine areas, providing a link with the open sea. There is also a fluctuating sediment load in the water coming into estuaries from both the rivers and offshore areas. Temporal changes are perhaps shown most dramatically by wading birds, many of which arrive only at particular seasons, but may have drastic effects on the infauna. Infauna are also affected by fluctuations in physical stresses such as erosion and deposition, as well as by the changing supply of phytoplankton and suspended detritus. While it is therefore possible—but not simple—to gain an overall understanding of ecosystems in each of the habitats we have discussed, it is much more difficult to form an overall view of a whole estuarine system. In addition, the descriptions of estuaries and lagoons in Chapter 7 should have made it very clear that almost every estuary is unique. How can we approach the idea of an estuary as a functional unit?

Food webs: who eats what?

One of the classical ways of assessing how species interact is to observe what they eat, and what in turn eats them. In this way, species can be placed in trophic levels, from primary producers via herbivores and detritivores to carnivores. The resulting pattern of linkages is referred to as a food web. For example, in many north temperate estuaries, food webs are thought to have detritus at their base, derived partly from local macrophytes and partly from material imported into the estuary (Fig. 10.1). The detritus is eaten by detritivores which may be either deposit feeders or suspension feeders, or both, while some of the original macrophytes are eaten by herbivores. The detritivores are then consumed by vertebrate and invertebrate predators such as fish, birds, and crabs. In parallel with this detritus-based chain is a phytoplankton-based chain that interacts with it.

In estuaries nearer to the tropics, the higher levels of food webs are dominated by a much greater diversity of fish than in temperate zones, ranging from

Fig. 10.1 Generalized food web for a north temperate estuary, showing major links between trophic levels. Most of the downward links to the detritus pool have been omitted for clarity.

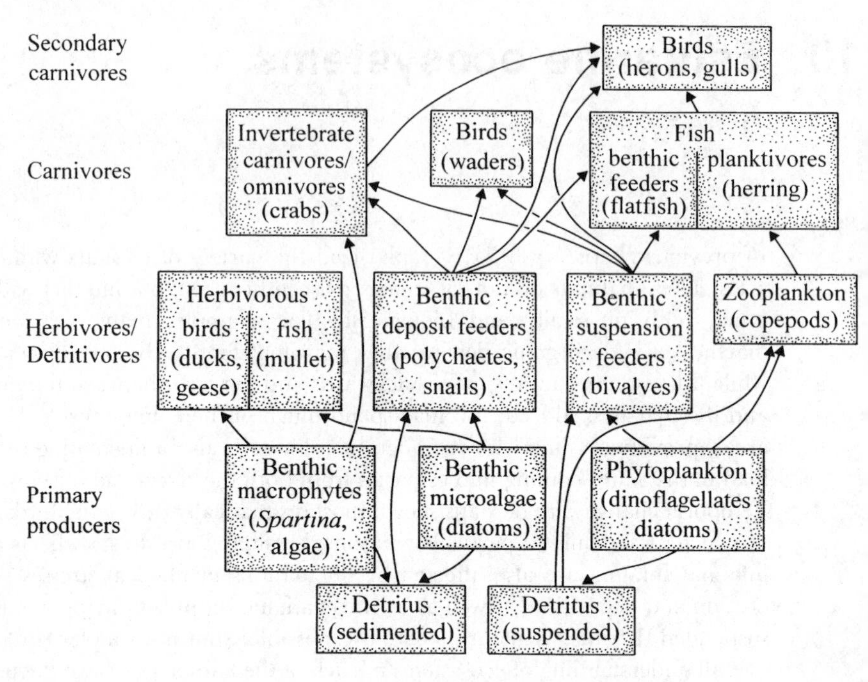

herbivores and mud feeders to benthos feeders, planktivores, and piscivores (Blaber, 1997). Depending upon physical conditions, the proportions of these various kinds of fish can vary greatly, as shown by a comparison of two coastal lakes in South Africa. Lake St Lucia is shallow and turbid, but in spite of its turbidity has a high planktonic productivity. It has many planktivorous fish and a smaller number of piscivores that feed on them. Lake St Lucia also has quite high numbers of fish that feed on the macrobenthos, and mud feeders such as the mullets. The higher trophic levels are thus supported both by phytoplankton production and by detritus (Fig. 10.2). Lake Nhlange, in contrast, has clear water but a low nutrient supply, and consequently little plankton production. The bulk of fish are made up of those eating the macrobenthos and the mud eaters, so here the food webs are primarily detritus based—planktivores account for only 2% of fish biomass, compared with 38% in Lake St Lucia.

Creation of such food webs involves an immense amount of observation: examination of gut contents must be integrated with observations of feeding behaviour of dominant organisms, and the whole considered over a period of at least a year. For detritus-based systems, the lower levels of such a food web are very difficult to sort out. Where does the detritus come from? As discussed in Chapter 4, do detritivores really eat mostly non-living detritus, or do they consume benthic micro-algae and bacteria? One technique that has recently aided discussion of some of these points is the use of stable carbon isotopes (see

Fig. 10.2 Comparison of food webs in two coastal lakes in South Africa, emphasizing trophic variety in the fish fauna (circles). The figures for percentage show % total fish biomass. Other trophic units are shown as rectangles. (After Blaber, 1997.)

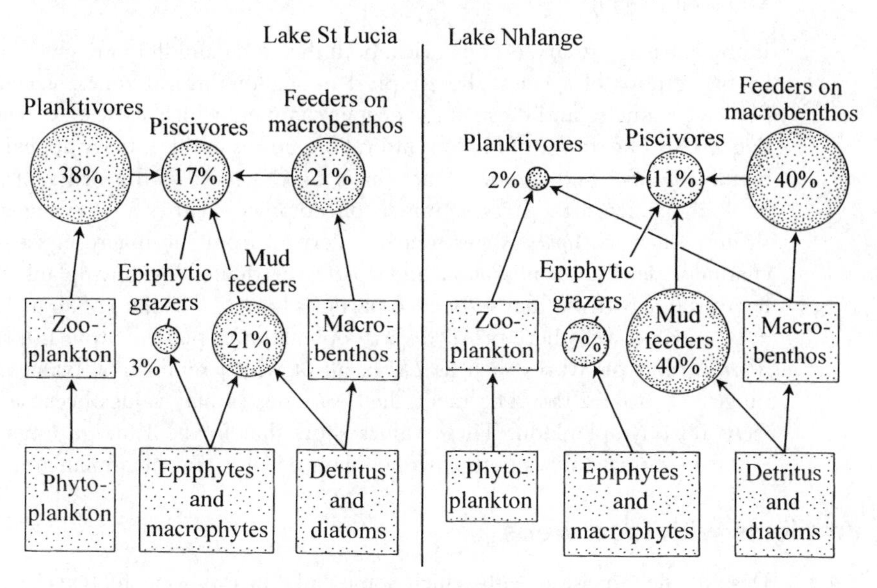

p. 89 and p. 112). Here we briefly discuss their use in evaluating estuarine food webs.

The use of stable carbon isotopes

Values of stable carbon isotopes vary with the carbon source, and can therefore be used to trace the origins of organic detritus. Terrestrial sources usually have depleted values of the isotope ^{13}C, i.e. very negative values when the $\delta^{13}C$ ratio is expressed as ‰. Marine algae, in contrast, usually show enrichment, i.e. they have less negative values of $\delta^{13}C$. River-borne detritus can therefore, in principle, be distinguished from that of marine origin simply by measuring $\delta^{13}C$. Care needs to be used, however, because many other factors affect the values of $\delta^{13}C$ in organisms. For example, the values in phytoplankton vary with growth rate as well as with carbon source. In the Tamagawa estuary, Japan, suspended particulate organic matter, derived from phytoplankton, can show $\delta^{13}C$ ratios varying from −30‰ at low growth to −15‰ at high growth rates in eutrophic regions (Ogawa and Ogura, 1997). In this case, isotope ratios in the estuary do not reflect merely a mixing of riverine and marine carbon sources, but variation in phytoplankton growth as well.

With this caution in mind, carbon and other isotopes can be very useful in investigating estuarine food chains. For example, in a Mississippi salt marsh, vascular plants such as *Spartina*, which have $\delta^{13}C$ values of about −13‰, are accompanied by benthic algae, e.g. *Vaucheria* and *Enteromorpha*, with $\delta^{13}C$ values of about −19‰. The animals in the marsh have values ranging from −22‰ to

−18‰. The similarity of these values to those of the algae suggests that their major food supply derives from the algae rather than from *Spartina* (Sullivan and Moncreiff, 1990).

In the Kariega estuary, South Africa, both the fauna and flora are characterized by two groups of species: those typical of shallow littoral zones, which show $\delta^{13}C$ enrichment, and those in the deeper channels, which show $\delta^{13}C$ depletion (Fig. 10.3). The source of food for littoral organisms in this estuary appears to be mainly *Spartina* and *Zostera*, which have $\delta^{13}C$ values of the order of −8 to −14‰. Mullet such as *Liza* browse on epiphytes with similar values, while shrimps such as *Alpheus* ingest detritus derived from the macrophytes. In the channels, detritus from *Zostera* and *Spartina* is diluted by phytoplankton and terrestrial detritus so that suspended particles have $\delta^{13}C$ values of about −20 to −24‰. Food webs here proceed via zooplankton to planktivorous fish such as *Atherina*, and piscivores such as *Lichia*, all of which retain $\delta^{13}C$ values in the range −17 to −20‰. The razor shell *Solen* has similar values because it also feeds on phytoplankton. These values show that in the Kariega estuary, the estuarine fauna utilizes both vascular plant detritus and phytoplankton.

Problems with food webs

Despite the precision with which some links in estuarine food webs can be characterized, there remain problems with an approach based on food webs as we have described them. One problem is that biologists usually build up webs

Fig. 10.3 Food webs in the Kariega estuary, South Africa, based on $\delta^{13}C$ isotope analysis. The food web in the shallow littoral zone depends mostly on $\delta^{13}C$-enriched material. The macrophyte *Sarcocornia* is depleted in $\delta^{13}C$, but is not an important food source. The food web in the deeper channels, including the suspension feeding bivalve *Solen*, depends mostly upon $\delta^{13}C$-depleted material. (After Paterson and Whitfield, 1997.)

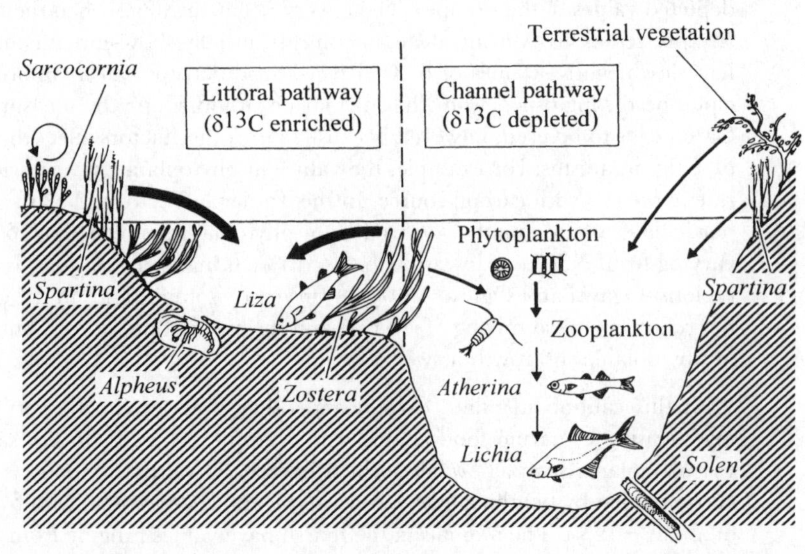

by fitting individual species—or more often groups of species—into trophic levels. Yet many organisms cannot be so constrained. In Fig. 10.1, for instance, crabs are mainly carnivores but also eat plant material so can be regarded as omnivores, and thus exist on two trophic levels. In addition, the links between species often become very tangled at higher trophic levels—many carnivores such as birds and fish feed on a great variety of prey. Understanding all these links is extremely time-consuming, even if it is possible. When it comes to quantification, the large number of links causes further difficulties. One way of expressing the relative importance of the various groups is to quantify their relative biomass, as in Fig. 10.2, but this ignores the fact that biomass at any one time (standing stock) itself depends upon a balance between production and consumption.

Nevertheless, food webs can give a convincing overall view of estuarine ecosystems. In the Ythan estuary, Scotland, a 30-year research programme has documented a web comprising 92 species, involving 409 links between them. This complex web effectively summarizes a series of food chains, some of which have as many as nine component links (Hall and Raffaelli, 1991). It is hard to think of a better way of viewing such a complex system.

Food webs containing large numbers of species are usually compressed by aggregating species into groups, and as this is done some information is inevitably lost. Does this affect the validity of the webs? In the Ythan, the length of individual food chains within the web diminishes as aggregation increases, because the intermediate species become incorporated in trophic levels above or below their real levels. Nevertheless, most properties of the original food web, such as the proportions of predators to prey, and the relationship between number of trophic links and number of species, stay unchanged, showing the web's robustness (Hall and Raffaelli, 1991).

Food webs from a variety of habitats in fact show remarkable similarities, despite the varying degrees to which species are aggregated into groups. For example, the average proportion of species at different trophic levels is relatively invariant (Pimm *et al.*, 1991). Despite their shortcomings, therefore, food webs are likely to remain extremely useful tools for investigating estuarine ecosystems.

Energy flow

Another way of approaching ecological relationships between species in a community is to consider the interactions in a food web in terms of energy flow. This approach has the advantage of defining the most important links, and of pointing out the 'dominant' organisms. It also often allows many intermediate links to be filled in approximately, if higher and lower trophic links have already been quantified. Although such undertakings are extremely time-consuming and labour intensive, they can give a more objective view than food webs.

We begin by taking as an example the Ythan estuary, already mentioned above. Here the transfer of energy is expressed in terms of carbon flow, and the species are aggregated into 10 quite large trophic groups. The simplified version in Fig. 10.4 ignores energy lost in respiration, faeces, and so on, and so ignores a large number of loops that feed back into the detritus pools. Bearing these simplifications in mind, we can follow the flow of carbon from producers through the consumer links.

There are several sources of carbon in the Ythan (Baird and Milne, 1981). First there is the allochthonous (i.e. imported) carbon from the River Ythan and from the sea, which feeds into the category of suspended particulate organic carbon (POC). Then there are the autochthonous sources (i.e. those produced within the estuary): the benthic macrophytes such as the alga *Enteromorpha*; phytoplankton, which produce only one-fifth as much; and benthic microflora such as the diatoms, which produce only one-tenth as much as the macrophytes. A large fraction of this production is probably exported from the estuary. Very little passes directly to the herbivores. Some macrophytes are eaten by birds such as wigeon, and some of the phytoplankton is eaten by zooplanktonic copepods and by the suspension feeder *Mytilus*. But the majority is thought to be channelled to a second POC category—that in the sediment. Sedimented POC then provides the basis for most of the benthic food chains in the Ythan, and indeed this is a general feature of most aquatic ecosystems (Mann, 1988).

Fig. 10.4 Energy flow in the Ythan estuary, Scotland. The width of the arrows is roughly in proportion to energy flow. Values are shown in gC/m^2 per year. POC, particulate organic carbon. (After Baird and Milne, 1981.)

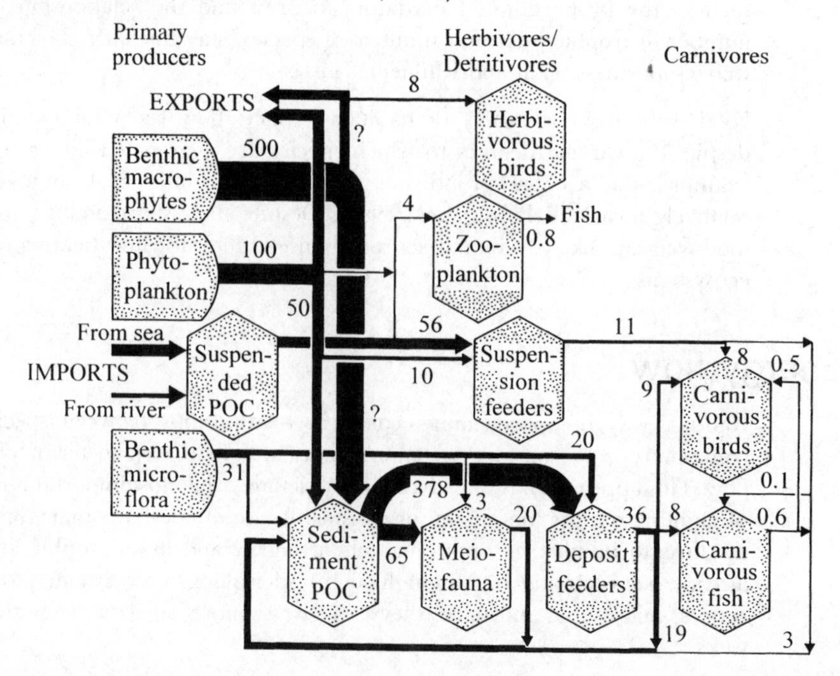

The benthos accounts for much greater energy flow than do pelagic systems. But within the benthos, there is great uncertainty about the fate of meiofaunal production—does it form its own enclosed food chains, or is a large proportion eaten by macrofauna? The macrofauna itself consists mostly of deposit feeding invertebrates, and these, as well as the suspension feeders such as *Mytilus*, are eaten by predatory birds and fish.

The overall picture of energy flow is thus clearly one in which the major routes are via the deposit feeding benthos, and not via the plankton. There are, however, several major points this energy web does not address. For instance, how does dissolved organic carbon (DOC) fit into this picture, and is the microbial loop important? How does the picture in Fig. 10.4 vary seasonally? What is the overall balance between imports and exports, i.e. does the Ythan act as a source or a sink for carbon? Some of these questions can be approached by considering energy flow in other estuarine systems.

In the Chesapeake Bay, the flow of carbon between 36 separate compartments has been analysed (Baird and Ulanowicz, 1989). Here the autochthonous production by phytoplankton dominates the sources of carbon, and benthic macrophytes are insignificant. A considerable fraction of phytoplanktonic production, perhaps 25%, is estimated to be lost as DOC. This material is utilized by bacteria, fuels the microbial loop, and hence supports zooplanktonic growth. Zooplankton also directly consume over 30% of phytoplankton production in spring, but this value falls to only 5% in summer, because then the zooplankton are, in turn, consumed by ctenophores and jellyfish: seasonal changes are of great importance. The majority of the phytoplanktonic production is not eaten by zooplankton, and so falls to the benthos and fuels the detritus-based food chains. The overall dependence upon this detrital material is well shown by comparing production by deposit feeders (about 33 gC/m^2 per year) and suspension feeders (about 5 gC/m^2 per year): in the Chesapeake, despite clearer water and higher phytoplanktonic production than the Ythan, most of the energy is still channelled via detritus.

A third example, the Kromme estuary in South Africa, is more like the Ythan in that it has little phytoplankton production, and its major carbon sources are benthic macrophytes. Here, however, salt marshes with *Spartina*, and mudflats with *Zostera* are present instead of algae. These decay to provide a detritus source, and soluble products also leach out and contribute to the microbial loop. Most of the DOC is, however, imported from the sea. Interchanges with the sea also drastically affect the phytoplankton: with a flushing time of 26 h, compared with 42 days in the Chesapeake Bay (Heymans and Baird, 1995), phytoplankton is rapidly flushed out to sea.

Exchange with the sea is evidently very important in the Kromme, and we go on now to consider in more detail how important such exchanges are in estuaries as a whole.

Imports and exports: exchange with the sea

Attempts to derive carbon 'budgets' for estuaries are fraught with difficulties. Enormous amounts of material pass into and out of any estuary on tidal, seasonal, and yearly scales. Any overall import or export thus reflects a very small percentage of these amounts, and is extremely difficult to measure. There is also a wide spectrum of estuarine types with different physical and biological characteristics that affect the overall balance. Those with extensive salt marshes may produce large quantities of detritus, but as we noted in Chapter 5, not all salt marshes export this material. Those with very turbid water may produce very little organic carbon by photosynthesis, while those with clear water have high phytoplankton production. Some estuaries appear to have sediment-trapping mechanisms, whereas others appear to export sediment. Two estimates indicate the range of possibilities (Wolff in Coull, 1977). First, Barataria Bay, Louisiana, which is fringed with salt marshes and has clear water in which phytoplankton and benthic plants flourish, may *export* particulate material to the extent of 750 gC/m^2 per year. In contrast, the Grevelingen estuary, the Netherlands, in which phytoplankton is the major biological carbon source, is estimated to *import* particulates up to 450 gC/m^2 per year.

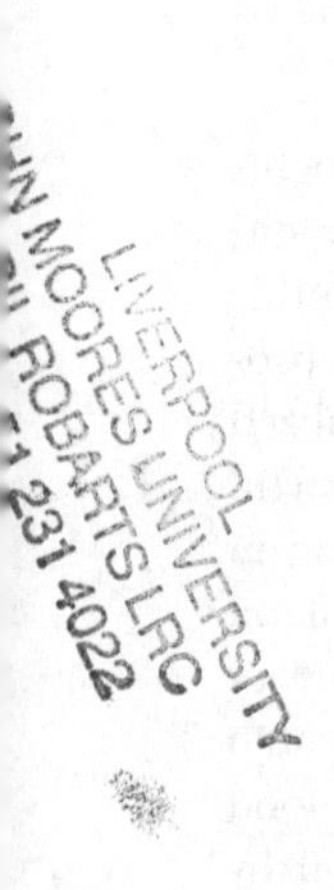

Besides this interchange of detrital carbon with the sea, estuaries exchange carbon dioxide with the air: in the Scheldt estuary, the Netherlands, about 60% of the respiratory carbon dioxide is released to the atmosphere, while the rest is sedimented or remains in solution (Gattuso *et al.*, 1998). There may also be substantial exchanges in the form of dissolved organic carbon, and in the form of living biomass. Fish and crustaceans mature in estuarine waters (Chapter 9) and their emigration then represents a loss of carbon. Similarly, visiting birds feed on the flats, and increase their biomass prior to migration. When they leave, they effectively export carbon. For example, in the Tagus estuary, Portugal, birds remove about 3000 kg (in dry weight) of benthic animals *per day*, during winter. In a year this amounts to more than 10% of benthic production, and values in other estuaries vary from 6 to 44% of production (Moreira, 1997).

How should we consider the overall carbon balance of an estuary? Are estuaries net consumers or net producers of organic matter? Despite the technical difficulties, the balance between 'system primary production' and 'system respiration' has now been established for more than 20 estuarine and shallow coastal systems (Heip *et al.*, 1995). In general, these systems consume more organic carbon than they produce—they are called heterotrophic by analogy with organisms that consume organic carbon. Heterotrophic systems are particularly common where conditions are turbid—such as the Westerschelde estuary and North San Francisco Bay. However, some shallow, clear estuaries such as the Texas bays and areas on the Georgia coast also show high degrees of carbon import. On the other hand, fjords and some negative estuaries in Australia produce more carbon than they consume—they are called autotrophic by analogy with plants that produce organic carbon.

The explanation for this distribution of heterotrophic and autotrophic systems may lie in the factors that control primary production (Heip *et al.*, 1995). When the degree of import or export of carbon is plotted against overall productivity, a pattern emerges (Fig. 10.5). In very turbid estuaries, light levels are low and primary production is in fact independent of net organic import. But when light levels are above a certain critical level, primary production rises in relation to import of organic carbon. Why should this rise occur? After all, phytoplankton do not normally take in any significant quantity of fixed carbon—they are autotrophs, and produce their own carbon compounds. The answer may be that estuaries recycle most of the nutrient elements like nitrogen and phosphorus that are bound up in the imported organic compounds. These elements are remineralized, i.e. their organic links are broken down so that compounds such as proteins release their nitrogen in the form of nitrates which can be utilized by the plants. Release of nutrients stimulates plant growth. The recycling of nutrients within estuaries may therefore prove to be one of the most critical factors in estuarine ecosystems.

Modelling and predictions

When sufficient is known about the flow of carbon in an estuary, and the interchange between the estuary, its river, and the sea, it is possible to construct some form of model representing it. Modelling is, however, a complex operation and usually has inherent inaccuracies. And models can only be as accurate

Fig. 10.5 Primary production in estuarine and shallow coastal systems in relation to net organic carbon exchange. The regression line is calculated only for points above a production rate of 160 gC/m^2 per year. Below this (stippled area), primary production is light limited. Open circles show fjords and an Australian negative estuary, which export carbon and are not light limited. (Modified after Heip *et al.*, 1995.)

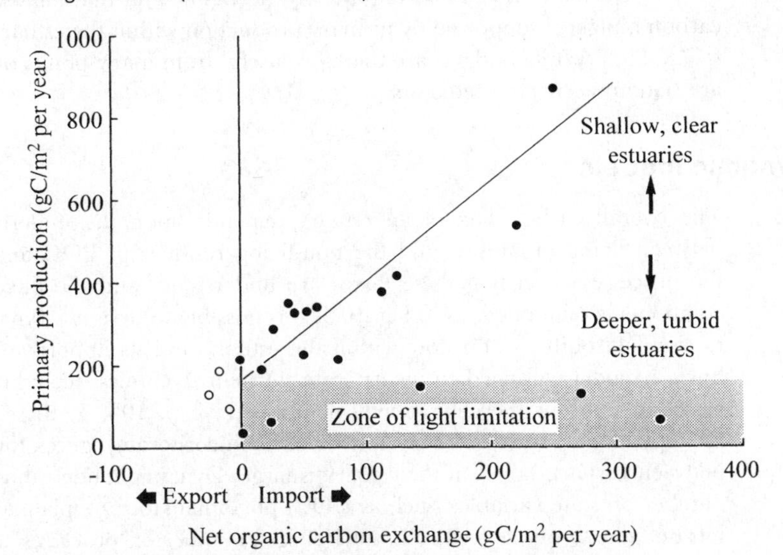

as the information that is fed into them. So why are such models useful? There are two answers for biologists, although modellers may think of more. First, making a model may help to increase our understanding of the ecosystem. If we can make a model that accurately mimics real features such as, say, fluctuations in primary production by phytoplankton or secondary production by fish, we may be further towards understanding the processes that underlie such fluctuations. If, on the other hand, initial models do not work accurately, the inconsistencies can be used to trigger more research to discover ways of modifying the model. The second use of models is to allow predictions of what might happen if conditions change. Many estuaries are threatened with 'development' in the form of barrages, landfill, and so on, as we shall discuss in Chapter 11. A model may help to predict the consequences of such developments.

Steady-state models

The simplest models consider the overall carbon budget of an estuary, as discussed in the last section. How much carbon is imported, how much exported, how much is fixed by photosynthesis and how much lost by respiration? These 'steady-state' models have, however, proved notoriously unreliable. Partly this is because, as we pointed out on p. 194, the overall carbon balance depends upon very small differences between very large variables: the error factor is therefore large. In addition, steady-state models are usually over-simplified when they attempt to consider the budget for an entire estuary over a period of, say, a year: the budget is affected by many dynamic factors that cannot be included in the model. In the Oosterschelde estuary, the Netherlands, different steady-state models suggest varying importance of phytoplankton production and detrital imports. While one calculates that primary production contributes only a third as much carbon as imports of POC, another calculates that POC imports are negligible and that almost the entire carbon budget is supported by primary production within the estuary (Scholten *et al.*, 1990). While budgets are therefore useful from many points of view, they are seldom accurate predictors.

Dynamic models

The overall carbon flux of an estuary depends upon the underlying fluxes between living organisms and the non-living residues of POC and DOC. If the processes governing these fluxes are understood, and the fluxes between components quantified, as in Fig. 10.4, it is possible to design a dynamic model to simulate both carbon flow within the estuary and its imports and exports. Such a model will need to incorporate nutrient dynamics, including the processes of nutrient recycling briefly mentioned on p. 195, as these will affect phytoplankton growth. It may also need to incorporate figures for transport between various regions if the estuary is large. Dynamic models thus contain a number of 'state variables' such as DOC, phytoplankton, zooplankton, silicate, nitrate, and so on, as well as a much larger number of 'processes' such as the rates of primary production and feeding that determine interchange between the variables.

Three examples give an idea of the use of such models. In the Oosterschelde, a dynamic model was developed to estimate the effect of a storm-surge barrier on overall production. Part of the interest here is that the Oosterschelde has important cockle fisheries and large areas of mussel culture, both depending upon phytoplankton for food. The model uses 11 state variables, including carbon pools and nutrients (Fig. 10.6). The main output of the model is in terms of organic carbon, so it can be used to determine whether there is sufficient food for suspension feeders at particular levels of shellfish exploitation. Overall, the model predicts that the storm-surge barrier will have remarkably little effect. Phytoplanktonic production *in situ* is regarded as being the main source of organic carbon, and this production will, if anything, rise with a barrier in place: the Oosterschelde is a self-sustaining system (Scholten *et al.*, 1990).

In the neighbouring Westerschelde estuary, in contrast, an ecosystem model shows that phytoplanktonic production is relatively low—only about one-tenth of that in the Oosterschelde. Here the system is markedly heterotrophic—but the imported carbon comes not from the sea, but from waste discharges. The 'model estuary' acts as a trap for reactive organic matter, but does also export some of the more refractory material to the sea (Soetaert and Herman, 1995).

A model of carbon flow in the Severn estuary was developed in relation to proposals for a tidal power barrage that would have greatly reduced tidal flow in and out of the estuary. This model involves 15 state variables and a much

Fig. 10.6 The major state variables used in a model of carbon flow in the Oosterschelde estuary, the Netherlands, together with the relations between them (arrows). Additional state variables used are salinity and dissolved oxygen. (After Scholten *et al.*, 1990.)

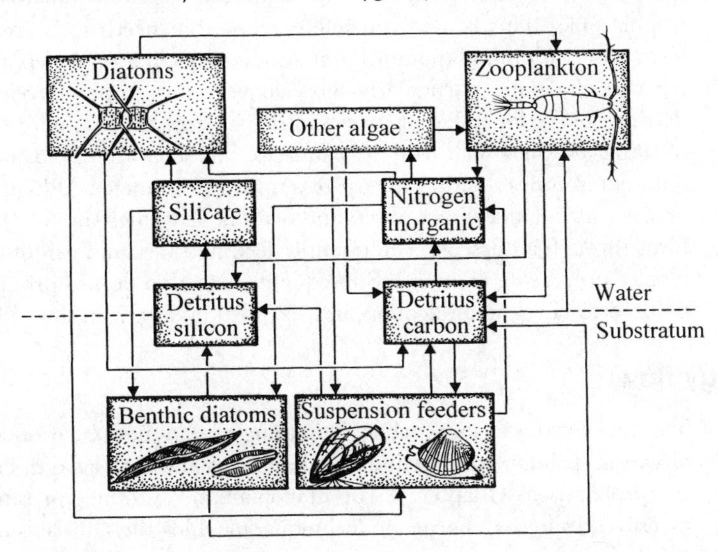

larger number of processes, and has successfully been used to simulate phytoplankton and zooplankton biomass over a period of 10 years (Radford, 1994). The model predicts that because a barrage would reduce turbidity in the water column, primary production would increase by a factor of over 50. Phytoplankton biomass would only double, however, because zooplankton and benthic suspension feeders would increase. In this case, the proposed barrage would therefore considerably modify the Severn ecosystem.

Models, of course, depend upon many assumptions, as it is impossible to measure all the processes accurately, or even to be sure that all relevant processes have been included. They therefore have a degree of inaccuracy, and it is important that attempts are made to quantify this. The longer a model can be tested against measurements in the 'real' estuary, the more the inaccuracy can be reduced. In the Oosterschelde, predictions were subjected to a stepwise 'uncertainty analysis' (Scholten *et al.*, 1990). This type of approach gives an indication of the reliability of the model, and is essential if models are to become more widely used.

Techniques

To establish the details of food webs, or the overall carbon balance of an estuary, requires time-consuming and specialist techniques. Most food webs can only be constructed with any confidence when research teams have investigated an area over a considerable time. Here we can only give a very broad brush outline of the techniques used.

Food webs

The construction of a basic food web can be attempted from observation of species present, gut contents, and a literature search aimed at establishing trophic links. This is a worthwhile exercise, but needs to be refined by long-term observation and quantification. Some steps in a food web are particularly hard to elucidate, particularly those at very low trophic levels: What is the detritus source for different species of detritivores, for example? Here the ratios of stable isotopes such as $\delta^{13}C$ can help, but separation and measurement of different isotopes requires the use of a mass spectrometer. At high trophic levels, on the other hand, direct observation can be extremely useful. Thus, watching birds through a telescope can quantify feeding rates, and can often identify the type of prey being captured. Subsequent inspection of bird droppings can then be used to check identification and size of prey (e.g. Moreira, 1997).

Energy flow

The first hurdle to cross when estimating energy flow is to obtain a measure of overall primary production. For benthic plants, we have discussed some of the problems in Chapter 5. For phytoplankton, production rates are usually measured using a radiocarbon technique, in which the sample is incubated with a small amount of radio-labelled bicarbonate, and uptake of ^{14}C is monitored. This, and other techniques such as the old 'light-and-dark-bottle' method (in which changes in oxygen concentrations are compared between clear bottles and darkened ones) do have inherent problems, however (Dring, 1982), and there are still arguments about accuracy.

In addition to *in situ* production, some estimate must be made of import and export, both of POC and DOC. Because of the overall movements of water, this is essentially an oceanographic problem (Heip *et al.*, 1995).

Movement of energy through the food web is usually followed as movement of carbon. Carbon budgets are constructed using the equation

$$C = P + R + E + U,$$

where C is consumption, P is secondary production, R is respiration, E is egestion (loss in faeces), and U is loss in urine. An example is given by Baird and Ulanowicz (1989).

Measurement of organic carbon must be preceded by using filters to separate the sample into a nominally dissolved fraction and a nominally particulate fraction, and by removal of inorganic carbon. Because most of the latter exists as carbonate, it can be removed by simple acidification. Carbon content can then be analysed chemically (see Tett in Baker and Wolff, 1987), but is now more usually analysed using an automated CHN analyser (Williams in Head, 1985; Kramer *et al.*, 1994).

11 Uses and abuses: human impacts and counter-measures

Because of their strategic position, offering sites for river crossings, harbours, and industries of many sorts, estuaries have become a focus for large proportions of the world's population. In the USA, the Chesapeake Bay has cities such as Richmond, Washington, and Baltimore at the head of its tributary subestuaries. In Britain, London and its docks dominate the Thames estuary, and more than 30% of the population lives around estuaries. In Australia, 60% of the population is concentrated in cities on the edge of estuaries. It is not, therefore, surprising that estuaries have been subject to enormous impact from human waste products. Indeed, they often appear to have been regarded as natural waste-disposal units in the mistaken belief that material emptied into them will be rapidly flushed out to sea and dispersed. Experiences over the last 150 years have shown that soft shores in general, and estuaries and lagoons in particular, are very susceptible to the effects of man-made waste. Sandy shores on open coasts have for the most part escaped from damage, because dilution is sufficient to prevent pollution. But on the finer substrata in sheltered areas, flushing time is usually extremely long so that pollutants can rapidly build up to problem levels.

In this chapter we discuss first some of the problems involved in demonstrating that damage is occurring. Then we go on to detail some of the major problems in estuaries. Lastly we discuss possible ways in which damage can be dealt with, or prevented.

Detecting damage—Has there been pollution?

In many estuaries it might seem inherently obvious that pollution is occurring: in other words, that substances (or energy) have been introduced by humans, resulting in deleterious effects. However, demonstrating that any specific contaminant or procedure has produced a change is far from easy. In an ideal situation, an environmental impact should be investigated by comparing populations or communities before the impact with those after it. Such comparisons involve taking account of natural variations in time and space, and thus necessitate sampling over a period of time and at a large number of sites. There should also be multiple control sites to allow for natural changes (Underwood in

Underwood and Chapman, 1995). This sort of study is essential for planned impacts such as the building of barrages, marinas, and so on, most of which have 'impact studies' built in to the original proposals. However, most pollution events are not planned, or if they are, they have been in existence for so long that controls are impossible. Unless a monitoring scheme happens to have been running in the area, biologists investigating the effects of estuarine pollution have to fall back on other techniques.

One approach is to measure the concentration of supposedly damaging substances in the field, and then to measure their toxicity to various species in the laboratory. For example, if the concentration of zinc in estuarine water is 10^{-2} M, and this concentration kills 50% of mussels in a laboratory tank within a few hours, zinc in the estuary is almost certainly important in determining mussel mortality. But there are many complicating factors. For instance, some heavy metals form organic complexes which can be more toxic than their inorganic salts, so it is important to know the form in which they exist at any site. There are also often synergistic effects—toxic effects of two factors together may cause *more* harm than the predicted sum of their individual effects. In all, toxicity tests are useful indicators, but they can seldom be used to predict the effects of an effluent in the natural environment.

Most other approaches involve analyses of community composition, in attempts to determine if this has changed from 'normal'. For instance, univariate analyses involve the calculation of a single index, such as species richness or a diversity index. Multivariate analyses involve comparisons of the composition of communities, and have sometimes been used successfully in combination with categories of 'indicator' species (Grall and Glémarec, 1997). But perhaps the most widespread methods are those based on the relationships between abundance and biomass. In undisturbed soft sediments, communities are usually dominated by relatively low numbers of individuals of many large-bodied species, while polluted sediments have high numbers of a few small-bodied species (Warwick, 1986). If the percentage dominance is plotted against ranked species numbers, the curves for biomass and numbers shift with increasing degrees of pollution (Fig. 11.1). Much of this response results from changes in composition of the polychaete fauna, and does not apply to molluscs, crustaceans, or echinoderms (Warwick and Clarke, 1994). These graphical methods, together with univariate and multivariate analyses of community composition, give a good indication of whether communities are 'natural' or 'polluted', although they do not indicate the cause of pollution.

The problems: impacts on estuarine and lagoonal ecosystems

The impacts caused by humans are so numerous that here we can only deal briefly with a few of them. The most widespread have in the past derived from the release of sewage into estuarine water, but threats to estuarine ecosystems

Fig. 11.1 Abundance–biomass comparisons (the 'ABC' method) for two hypothetical communities. In the unpolluted community, there is a high diversity (but few individuals) of large species. In the polluted community, there is a low diversity (but many individuals) of small species, particularly polychaetes. (After Warwick, 1986 and Warwick and Clarke, 1994.)

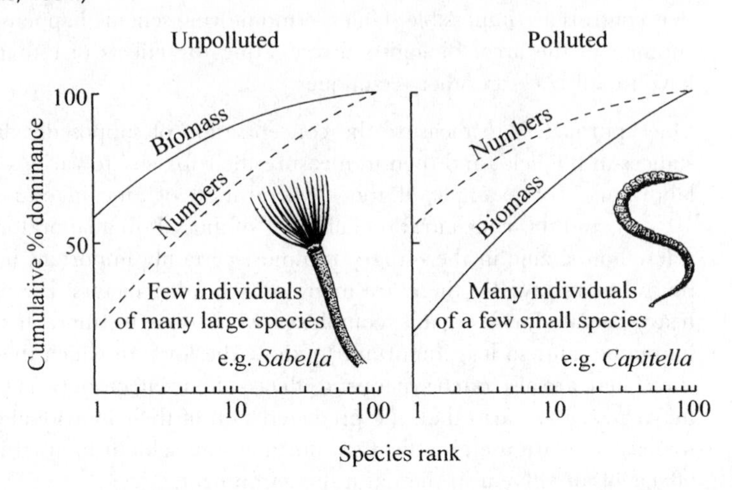

are now generated by many human activities, from industry and fisheries to reclamation and recreation. Enormous areas may be polluted by industrial effluents which contain not just one pollutant but a cocktail of damaging substances. Estuarine flats can be damaged merely by disturbance from visitors which prevents waders from feeding or roosting. Small areas such as lagoons are easily damaged by local problems such as rubbish tipping.

Sewage release, eutrophication, and oxygen shortage

The major harmful constituent of untreated sewage is organic matter, because when this is released into estuarine waters, bacterial oxidation absorbs large quantities of oxygen. Most rivers and estuaries in industrialized countries thus show some degree of 'oxygen sag' as already discussed on p. 144 (Fig. 7.7). However, treatment plants that remove much of the organic matter have greatly improved the situation, as we shall see (p. 215).

Organic loads, and reduced oxygen tensions, produce distinctive soft-sediment fauna, particularly encouraging opportunist species (Pearson and Rosenberg, 1978). In estuaries, these are often oligochaete worms. In the Thames, for example, *Tubifex* spp. were so abundant in polluted waters in the early 1970s that they were exploited as a food source by fish such as bream and waders such as dunlin. In the later 1970s, *Tubifex* densities declined from values as high as 300 000/m^2 to only 4000/m^2, as water quality improved (Wheeler, 1979).

When organic loads are high enough to produce complete anoxia, both pelagic and benthic communities die. For example, in the Tees estuary, which runs through industrial areas in the north of England, many upstream stations contained no benthic macrofauna in 1979, although by 1985 some recovery had been seen (Shillabeer and Tapp in McLusky *et al.*, 1990). These upstream stations contained both oligochaetes and the pollution-tolerant polychaete *Capitella capitata*. Fish also die out in total anoxia. A large stretch of the tidal Thames had no fish for many years during and after the industrial revolution, despite the fact that it was well known as a salmon river in the early nineteenth century. Even a small reduction in oxygen tension can reduce the number of fish species: the Tongati estuary in South Africa receives sewage effluent which reduces oxygen levels, and here the fish populations are impoverished. Mullet species, which are relatively tolerant of low oxygen, dominate (Blaber, 1997). In fact the tolerance of mullet is exploited in some Indian estuaries, where they are cultured using sewage effluent.

In many estuaries, the effects of sewage are spread seawards because sewage sludge from treatment plants is taken out to sea and dumped. For 60 years, until 1986, sludge from New York's treatment plants was dumped in the New York Bight at a depth of about 50 m. This procedure involved shipment of about 10 million tonnes/year, and caused organic enrichment at the dumping sites. Movement of the dump site to nearly 200 km offshore in 1986 ensured better dispersion, but dumping was discontinued in 1991, as it has been at many sites (Clark, 1997).

Even with the use of sewage treatment plants, release of sewage effluent into estuaries can cause problems. These arise because most plants provide only what are called primary and secondary treatments, which remove organic matter, but leave in solution high concentrations of nutrients such as nitrates and phosphates. In many estuaries additional nutrients are derived from agricultural fertilizers, which are dissolved in run-off from the land. Thus many estuaries are now eutrophic: the nutrients stimulate growth of both macroalgae such as *Enteromorpha* and *Ulva*, and microalgae such as diatoms. Decay of massive amounts of plant tissues then causes deoxygenation, which is now widespread (Diaz and Rosenberg, 1995). But despite the general acceptance of this picture of eutrophication, the correlation between increase in nutrients and degree of eutrophication is seldom clear cut (Gray, J.S. 1992). In the Baltic, nutrients have increased over the last 30 years (Fig. 11.2), but eutrophication has only occurred in local patches. Evidently the relationships between nutrient concentrations and plant growth require further investigation.

Excessive growth of algae can cause many other problems than those mediated via oxygen depletion. High nutrient levels have been linked to the occurrence of algal blooms known as red tides, in which dinoflagellates reach very high densities. The toxins they release may then result in fish kills. High nutrients are also involved in the formation of benthic macroalgal blooms, which are now common in estuaries and lagoons and on intertidal flats generally. These blooms outcompete seagrasses and other seaweeds, and may dramatically change the benthos in the underlying sediments. Their exact causes, however, remain unclear (Raffaelli *et al.*, 1998).

Fig. 11.2 Phosphate concentrations in part of the Baltic Sea (the Gotland Deep) over four decades. (After Gray, J.S., 1992.)

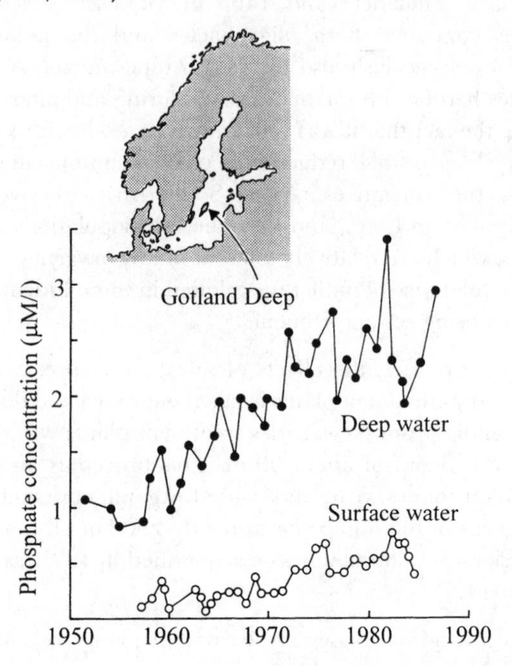

Chronic contamination and bioaccumulation

Industrial effluents and urban run-off contain an enormous variety of substances in addition to the faecal-derived organic matter and nutrients of sewage. A high proportion—perhaps 40%—of the oil that reaches the world's oceans is derived this way (Clark, 1997). Two other categories that are particularly important for estuaries and soft sediments are the halogenated hydrocarbons [such as the pesticide DDT and polychlorinated biphenyls (PCBs)] and heavy metals (such as zinc, cadmium, lead and mercury). These substances are not readily broken down in the natural environment—they are 'conservative' pollutants—and many tend to be taken up by organisms but are not excreted: they show the phenomenon of bioaccumulation, and may build up to increasingly higher concentrations at higher levels in food chains. In the Weser estuary, Germany, the cockle *Cerastoderma*, the lugworm *Arenicola* and the shrimp *Crangon* have concentrations of PCBs between 10 and 60 ng/g; while the sole *Solea*, which feeds on the benthos, has concentrations as high as 200 ng/g.

What effects do such concentrations have upon the bioaccumulating organisms? Answers are not easy to provide. For a start, halogenated hydrocarbons have a very low solubility in water, and are stored mainly in fat reserves: they may produce a harmful effect only when these reserves are mobilized, as for instance when the animal is short of food (Clark, 1997). However, both DDT and PCBs have been shown to reduce primary production in phytoplankton at very low concentrations, and fish are very sensitive to DDT in laboratory tests. Seabirds such as guillemots and pelicans can accumulate PCBs, DDT, and other pesticides, and declines in population or 'wrecks' in which large numbers die in a short time have been attributed to these accumulations. In almost every case, though, proof of the cause of death is lacking. Changes in food supply complicate the issue, and the concentrations of the pesticides within dead birds are highly variable. When steps are taken to reduce pesticide concentrations, the birds are indeed often seen to recover. For example, when DDT production was curtailed in the USA, the brown pelican colonies in California started to breed again after a gap of nearly 10 years. But even here, there are doubts about the cause of recovery because variations in sardine stocks, on which the pelicans feed, probably contributed to lack of breeding success.

In the case of the heavy metals, there seems even less conclusive evidence of their harmful effects, despite the fact that they can occur in high concentrations in estuarine sediments. Concentrations of zinc and copper in muddy sediments of British estuaries range from 30 to over 150 mg/kg, and lead from 30 to 70 mg/kg (Little and Smith, 1994). The concentrations of heavy metals in solution probably vary more widely than those in sediments, and many studies have utilized uptake by organisms to monitor overall exposure to the metals. For example, the alga *Enteromorpha* can be used as a convenient detector for zinc, cadmium, mercury, and lead (Say in McLusky *et al.*, 1990).

One reason for the apparent tolerance of fauna to heavy metals is their ability to sequester metals in granule form. *Mytilus*, for example, stores lead in granules in its digestive gland, while oysters can take up copper until it is more than 7000 times as concentrated as in sea water. They may literally turn green in the process, and have to be cleansed in uncontaminated water for a year before being sold, but the copper does not appear to harm them. Some metals do, however, have extensive sublethal effects—the example of tributyltin (TBT), used in anti-fouling paints, being perhaps the best known. Very small quantities of TBT cause changes to reproductive systems of molluscs and can reduce populations of shellfish. In the oyster *Crassostrea gigas*, TBT also causes reduced growth of the tissues, but excessive growth of the shell, so the oysters never grow large enough for sale. TBT paint has now been banned on small boats in much of the world, although it is still used on large ships.

Nevertheless, without controlled experiments of the type discussed earlier, it will in general be very difficult to be certain of the effects of any of the conservative pollutants.

Chronic contamination and catastrophic spills: the effects of oil

Because of the siting of ports and oil refineries in estuaries, many estuarine shores are subjected to low-level chronic fouling with oil. Oil may be released in refinery effluents and during discharge from tankers, in addition to the oil in other domestic wastes and in domestic sewage. What effects do these low-level discharges have on soft sediments and their biota? Oil refinery effluents contain, as well as oil, high levels of chemicals such as sulphides, phenols, and ammonia, so the relative effects of various components in long-term discharges are not easy to assess. The overall effect can, however, be considerable. Light doses of oil often actually stimulate plant growth—the grasses *Festuca* and *Puccinellia*, for instance, show increased shoot lengths after experimental addition of Kuwait crude oil. This observation emphasizes the point that addition of substances regarded as 'pollutants' does not always result in degradation. But the stimulatory effect does not occur with *Spartina*. Indeed, a salt marsh in Southampton Water, UK, showed an almost complete decline of *Spartina anglica* over a period of 20 years during which two refinery outfalls spread a light film of oil across the area. The effect on *Spartina* probably occurs because oil reduces the rate of gas exchange across the leaves; but it also promotes anoxic conditions in the mud and hence encourages die-back. When the oil content of refinery effluent was reduced from 25 p. p. m. to 14 p. p. m., *Spartina* gradually began to recolonize the marsh, and *Salicornia* recolonized large areas within 5 years (Fig. 11.3).

The effects of catastrophic oil spills are more dramatic: large numbers of sea birds may be killed by the oil, shellfish can be killed or tainted, and fish such as flounders that live in shallow waters where oil accumulates may also suffer immediate kills. But oil spills, like chronic contamination, also have long-term effects. For example, the tanker Amoco Cadiz, which grounded off Brittany, northwest France, in 1978, released about 223 000 tonnes of crude oil and about 64 000 tonnes was estimated to have contaminated the shoreline over a period of about a month. The oil drifted into small estuaries, bays and salt marshes and became incorporated into the sediments. Because the trapped oil continued to leach out of the sediments it caused renewed contamination for years. In the nearby Grande Ile salt marshes, some of which were heavily fouled,

Fig. 11.3 Recolonization of a salt marsh in Southampton Water, UK, following a reduction of oil content in two refinery outfalls. Reduction in oil occurred between 1972 and 1974, and the figure shows recolonization by 1977. (After Baker in Jefferies and Davy, 1979.)

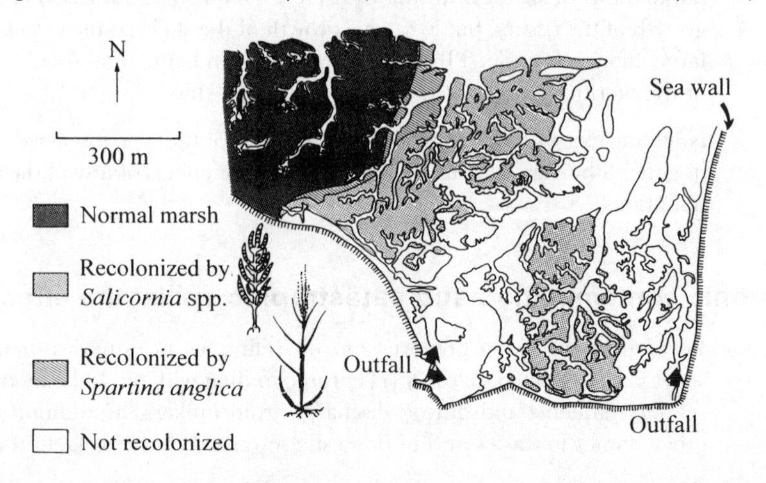

hydrocarbon levels remained above background at some sites 13 years later. While much of the oil was washed out into the sea, some was broken down by bacterial action. But many compounds, the so-called 'biomarkers', are very stable, and these remained unaltered even after 13 years (Mille *et al.*, 1998).

Oil spills can thus leave very long legacies, but in fact most biota recover quite quickly, provided the ecosystem is left to recover naturally. Far worse than the effects of the oil are the effects of emulsifiers and detergents that are sometimes used to remove it. Those used on the Torrey Canyon spill in Britain in 1967 are a classic case: communities on rocky shores that had been treated chemically took as much as 10 years to recover, and even then the rarer species had not recolonized, so overall diversity had not returned to its original levels. On soft shores, emulsifiers sprayed on to stranded oil allow it to penetrate downwards into the sediment where it may remain for a considerable time. Eventually it is decomposed by bacteria, but the decay creates anaerobic layers and may result in death of the infauna. Overall, the lesson is that the biota recover best from an oil spill without further human interference.

Physical alterations: the effects of land-claim

Estuarine shorelines, and particularly those dominated by salt marshes, have long been regarded as prime targets for land 'reclamation'—more appropriately known as land-claim. The consequent loss of marsh and mudflat has been eloquently mourned by Teal and Teal (1969). Around Boston, USA, an initial 480 ha of dry land available when the city was founded in 1630 was subsequently increased by draining and filling of 830 ha of salt marsh, mudflat, and sand flat so that by the beginning of the twentieth century all trace of the old marshes and flats had gone. In the Netherlands, where much of the country lies below 50 m, the intensity of land-claim has been much higher. In 1979, about 1.5 million ha of land, out of a total for the country of 3.4 million ha, consisted of land claimed from the sea (Knights, 1979). Much of this land is so low that it has to be drained by pumps.

In Britain, land-claim began as a serious measure in Roman times. Sea walls were built to enclose salt marshes for grazing, and progressive enclosure has subsequently extended these areas for a variety of uses from agriculture to industry. For example, in the Tees estuary, land-claim related to industry and the development of the port has removed all but a few per cent of the once large intertidal area (Fig. 11.4). As late as 1850, the Tees retained about 2740 ha of mudflats and salt marsh. By the mid-1970s this had been reduced to 470 ha (Davidson *et al.*, 1991).

What overall effect do such land-claims have upon the biology of estuaries? Losses of coastal and estuarine wetland have been estimated to lie between 25 and 50% of the total in both Britain and the USA so far (Davidson *et al.* in Jones, 1995), and land-claims continue. Such reductions in intertidal area have two major effects. First, they reduce the volume of water that the estuary can

Fig. 11.4 Land-claim in the Tees estuary, UK. Dates show the periods during which land-claim occurred. (After Davidson *et al.*, 1991.)

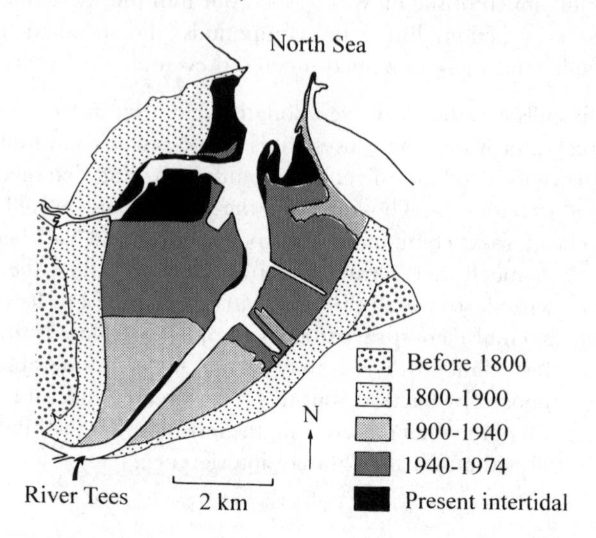

accommodate, and so alter the tidal regime. This type of physical change may be important, but has generally been ignored. Second, they reduce the overall biomass and production of benthic plants and animals, and hence affect production by the whole estuarine food web. There will be knock-on effects on the usefulness of the estuary as a nursery ground for fish, and on the ability of the mudflats and salt marshes to provide feeding grounds for birds. Much discussion has centred on the latter problem.

Waders and wildfowl use estuarine sites mainly when overwintering and when migrating to and from their breeding grounds. Often they need to build up fat reserves rapidly to prepare for migration. What effect does a reduction in their feeding grounds have on this process? Land-claim usually occurs very patchily (Fig. 11.4), so the first effect on feeding grounds is that they are fragmented. Because birds are easily disturbed, this fragmentation has an effect that is out of all proportion to the area claimed: birds are restricted to the centres of unclaimed patches. Within these patches, birds compete for prey, and as bird density increases, so does the rate of competition. Oystercatchers, for example, attack each other to steal mussels, and individuals that fail to obtain enough prey must emigrate to another mussel bed or face the threat of starvation. Other species such as redshank also interact, but indirectly: infaunal prey retreat into their burrows as redshanks walk over the substratum (Fig. 11.5), so food availability declines when redshank density increases. For any feeding area, as bird density increases it becomes harder for an individual to maintain its feeding rate. A balance is then reached between birds arriving to feed, and birds leaving to find other feeding grounds. The area has then reached its apparent 'carrying capacity', although it may be that if conditions elsewhere were to deteriorate, the carrying capacity of a particular area might increase. Defining

Fig. 11.5 The effect of a wader, the redshank *Tringa totanus*, on the appearance at the surface of a prey species, the amphipod *Corophium*. It takes 5–10 min for numbers of prey at the surface to recover. (After Gray, A., 1992.)

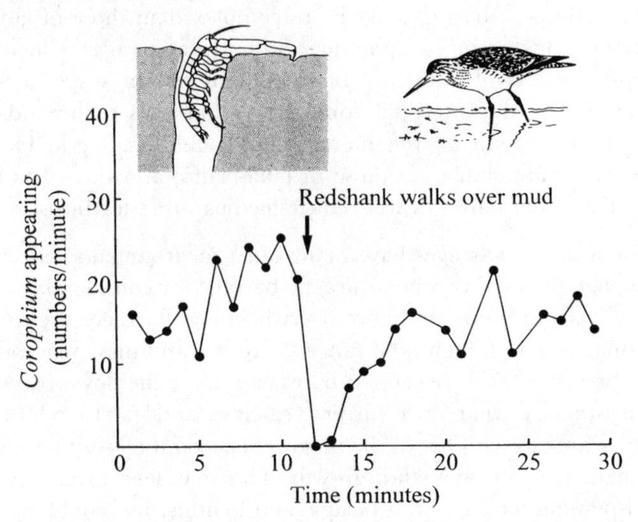

carrying capacity in absolute terms has so far proved elusive (Goss-Custard in Jones, 1995).

In spite of this uncertainty, the concept of carrying capacity is important because if the area of a particular feeding ground is diminished by, for example, land-claim, its carrying capacity will probably decrease. In theory, birds can then migrate to another mudflat, or another estuary. But in many countries, land-claim is affecting virtually *all* estuaries, so there is a danger that overall carrying capacity is being reduced. In this case, the population size of a species as a whole may be affected, as birds are forced to use less and less favoured areas, where food is in shorter supply and chances of survival are lower. As the population size will also be affected by conditions on the breeding grounds, however, assessment of the critical factors that determine shorebird numbers requires investigation on both local and global scales. As yet it has not been possible to integrate sufficient of these factors to predict the effects of loss of feeding grounds on bird populations, but evidently there is a great need for such predictions.

Physical alterations: the effects of barrages

Barrages across estuaries have been proposed for many sites and for many reasons, although few have so far been built. The first tidal power barrage was built at La Rance (northern France) in 1966, others in China and Russia, and one at Annapolis Royal in the Bay of Fundy (Canada). Proposed power barrages in the Minas Basin (Bay of Fundy), Severn (UK), and Mersey (UK) have given rise to much work on their predicted effects (Gray, A. 1992). Barrages have also been built to protect against storm surges (Oosterschelde, the

Netherlands), to impound fresh water for storage, and for recreational/urbanization purposes.

The effects of barrages are more complex than those of land-claim, and vary immensely depending upon their type. Most barrages reduce the intertidal area upstream. While amenity barrages may retain water as a pond, virtually eliminating the intertidal zone, power barrages reduce tidal rise and fall by up to 50%, reducing the intertidal but not eliminating it. Thus far the effects of barrages are similar to those of land-claim, and there has been considerable concern that barrages will reduce feeding areas for birds.

Tidal power barrages have many other environmental effects. They may even change the tidal regime *outside* the barrage for considerable distances, although the changes inside are more overwhelming. Barrages are usually proposed for estuaries with a high tidal range (5–16 m), and these macrotidal estuaries tend to be very turbid. Because a barrage reduces the flow of water into and out of an estuary, it will reduce current velocities, and lead to sedimentation and hence to clearer water. One of the major effects on the ecosystem will therefore be to encourage phytoplankton growth. This may lead to enhanced populations of suspension feeders, both pelagic and benthic, but could also in extreme cases lead to eutrophication. As the reduced current velocities will encourage stratification, increased surface production could lead to organic build-up and anoxia in near-bottom areas, and a consequent change in community structure. In addition, the reduced currents themselves will alter shear stress, so benthic communities may change drastically.

Barrages may also have significant effects on salt marshes. Decreased tidal range will lead to increased sedimentation, as we have just seen. If the sediment is deposited in the intertidal zone it may lead to downshore spread of the marshes. On the other hand, tidal power barrages will also allow increased wave action at particular levels on the shore because the water upstream does not rise and fall in a sine wave like a normal tide. Instead, it is held at a high level for a period after high tide in the sea, to allow a difference in height of water to develop across the barrage (Fig. 11.6). Only then is the water allowed to flow out through the turbines, generating power. The increased wave action high on the shore may cause erosion of the salt marsh. Accurate predictions of the balance between erosion and accretion are critical in determining the fate of salt marshes in barrage regimes.

The effects of barrages on fish also need to be taken into account. The physical effects of turbines on fish passing through them can be severe (Dadswell and Rulifson, 1994), and as we have discussed (p. 185), many fish migrate in and out of an estuary on a seasonal basis. In addition, juveniles that use estuaries as nurseries emigrate as recruits to the adult populations in the sea, while both anadromous and catadromous species pass through at some stage in their lives. Would a barrage have significant effects on these fish?

Fig. 11.6 Theoretical change in tide curves caused by a tidal power barrage operating on the ebb tide. Outside the barrage, the curve remains sinusoidal. In the basin landward of the barrage ('inside'), the amplitude of the curve is reduced and the water level is maintained at a high level for some time after the time of high water outside. Much of the intertidal zone becomes permanently submerged. (After Gray, A., 1992.)

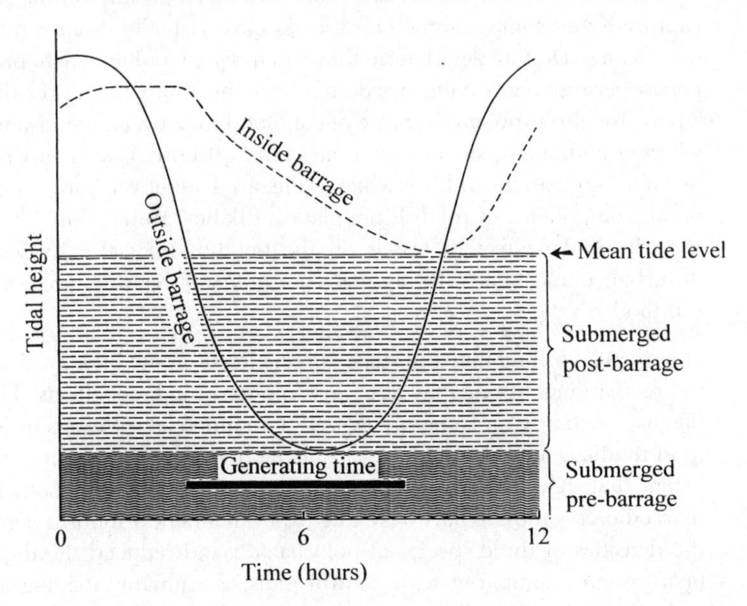

Emigrating fish need some sort of 'fish pass' to allow them to bypass the turbines safely, but in many types of barrage, passes are not practicable. It therefore seems likely that some proposed barrages would prevent recruitment from occurring. However, recruitment from any individual estuary is small in percentage terms, so loss of the entire recruitment from one estuary would probably not have any effect on marine stocks. It should be remembered, though, that the situation is analogous to that for birds: if many estuaries are affected, overall recruitment could be reduced. The same can be said for catadromous species such as eels, but the situation is more serious for anadromous species like salmon. These species must be able to emigrate without serious mortality if the stock in any particular river is to survive. At present insufficient is known about salmon behaviour to ensure that they could successfully bypass barrages.

Some effects of fisheries and aquaculture

In the temperate zone, many traditional estuarine fisheries are dwindling—such small-scale operations require much input of effort for very little profit, and are carried out mainly on an amateur basis. For instance, the Severn estuary used to have many fixed-net fisheries on its shoreline, dependent either on basket-work nets or on temporarily anchored rope nets, but these have virtually disappeared apart from some fixed nets still used for salmon. Some estuaries, such as the

Humber (UK) and the Chesapeake Bay (USA) do still have commercial fisheries for marine species, and in the tropics there are still many estuarine fisheries. What impact do these have on fish populations?

In many areas of Southeast Asia, there have been declines in the size of the fish captured, and some commercial species have virtually disappeared because of overfishing. Despite legislation that attempts to reduce such problems, they persist because many countries do not have the ability to enforce the legislation. There are also problems because of competition between the inshore fishermen, who use equipment such as seine nets, and offshore fishermen who use trawls: juveniles are caught inshore while adults are caught offshore, so causing stress on all components of the fish population (Blaber, 1997). While it has not been quantified, the effect of trawls on the benthos may also be severe, causing disturbance and hence decreasing the amount of benthic biomass available as fish food.

Commercial fishing may thus have direct effects on the biology of soft substrata, but recreational fishing may also have important indirect effects. This is because digging for bait is in some areas itself a commercial industry. In Maine, USA, most mudflats are dug or raked regularly, often several times in a summer, to an extent that they look like agricultural farms (Brown and Wilson, 1997). When marked plots on these flats were dug experimentally, 3 months digging reduced the densities of three species of polychaetes, and reduced the total number of taxa present, compared with control plots. Presumably digging acts as a disturbance, like those discussed for sublittoral sediments on p. 124. Similarly, commercial dredging disturbs the sediment, and depending upon substratum type, this may take years to recover (Newell *et al.*, 1998).

With declining fish stocks around the world, aquaculture has boomed, most notably in estuarine situations and fjords. In the tropics, enormous areas of mangroves have been converted to prawn culture. Thailand, for instance, lost 50% of its mangroves between 1985 and 1995 (Blaber, 1997), and the development of aquaculture ponds has been widespread in many countries of Southeast Asia. In the temperate zone, culture of fish in cages has resulted in other problems for estuarine habitats. Salmon have for many years been farmed commercially in floating cages in estuaries and inlets of the UK, Norway, and Tasmania. In areas where the flow of water is restricted, the wastes and uneaten food from such farms lead to de-oxygenation beneath the cages, and growth of pollution-indicator species such as *Capitella capitata* (Beveridge *et al.* in Ormond *et al.*, 1997). Pesticides used to keep the fish clear of parasites may also accumulate. For instance, ivermectin is widely used to combat parasitization by sea lice. Ivermectin is extremely toxic to many marine invertebrates, and there are concerns that it may cause ecological damage to the benthos in sheltered coastal inlets (Grant and Briggs, 1998). Effluents from rivers and lochs with fish farms may also change the characteristics of natural river water to such a degree that they deter native fish from using the area.

The impact of 'alien' species

Estuarine and marine communities throughout the world now contain large numbers of species transported by humans—the so-called 'aliens'. In the USA there are estimated to be at least 400 of these; in the Mediterranean, 240; in Australia, 70; and in the UK about 50, although these are almost certainly underestimates (Ruiz *et al.*, 1997). Estuaries and shallow-water muddy sediments have proportionately more of these aliens than rocky shores and open-coast sandy shores. For example, San Francisco Bay has 212 recorded alien species, whereas nearby outer coasts have less than 10. Mainly this difference probably arises because most imports are made, intentionally or not, into estuaries.

Taking San Francisco Bay as an example, the earliest aliens probably arrived before 1870 as fouling organisms on the bottoms of ships. In addition, from 1870 to the early twentieth century, the oyster *Crassostrea virginica* was imported from east-coast USA, bringing with it many associated species. From 1930 to 1960 the Pacific oyster *Crassostrea gigas* was imported from Japan, again with associated species. At the present day, the largest source of aliens is probably ballast water released from ships coming from overseas. Modern ships hold enormous volumes of ballast water. Because their journey times are now short, the survival rate of the organisms they carry has increased, and the rate at which they have contributed to invasions has increased over the past few decades.

What effects do alien species have on native communities? Initially, the addition of new species can be considered to increase biodiversity. If species take over empty niches, this may be thought of as advantageous. But although the effects of aliens on community structure have seldom been investigated in detail, some at least have altered native communities profoundly. One example is the invasion of San Francisco Bay by the Asian clam *Potamocorbula amurensis* (Fig. 11.7). Prior to 1987, the composition of the benthos varied in response to river flow. When flow was high, very few species were present, but in years of low flow, species typical of high salinity moved in. The communities were then dominated by the bivalves *Macoma*, *Mya*, and *Corbicula*. In 1987, one site was invaded by *Potamocorbula*, whose population density rose to more than 12 000 individuals/m^2, and in subsequent dry periods, native estuarine species failed to re-establish. The benthic community may have been permanently altered, although it should be noted that invading species do not always maintain the high densities achieved when they first arrive.

Long-term changes

In the long term, individual estuaries and the areas of sediment within them are ephemeral. Changes in sea level in the past have greatly affected estuarine outlines, and most present-day estuaries have only been relatively stable for the last 6000 years. Since this time, most have shown considerable infilling. Current forecasts that sea level will rise by perhaps 60 cm in the next 100 years must

Fig. 11.7 Changes in biomass of molluscs at one site in San Francisco Bay. Before 1987, native species of *Macoma, Mya,* and *Corbicula* were dominant. After the advent of the Asian clam *Potamocorbula amurensis,* these species have failed to re-establish, while biomass of *Potamocorbula* has been high. Bars show SD. (After Nichols *et al.,* 1990.)

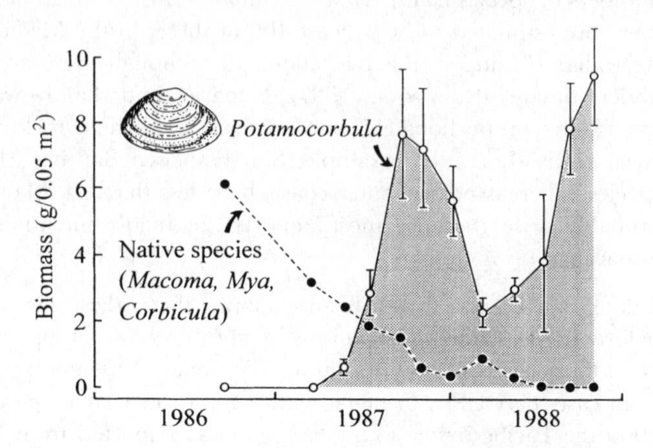

therefore be seen in this context. Nevertheless, climate change and sea-level rise *will* have immense and rapid effects on estuaries and other low-lying coasts (Holligan and Reiners, 1992).

Climate warming caused by the greenhouse effect is predicted to be greatest at high latitudes, but many warm-water species will probably expand their range both north and south from the equator. Mangroves, for example, will presumably expand their latitudinal range. Meanwhile, rising sea level will have drastic effects particularly within estuaries. Salt marshes provide an example of the complexities involved in forecasting such effects and in reacting to them. Rising sea level will inundate marshes and move the high-water mark nearer to artificial defences such as sea walls—the marsh will therefore become narrower unless its lower parts can grow to keep pace with the rising water. In eastern USA, marshes have so far kept pace with rising sea level because accretion rates are high (Adam, 1990). But where there is a shortage of sediment supply, marshes will diminish. Partly this is a consequence of human attempts to maintain fixed boundaries at the sea's edge: if it were not for sea defences, marshes would merely move inland as water levels rise.

The possibility of moving salt marshes brings us to a consideration of whether maintenance of the present lines of sea defence is desirable, or indeed even feasible. There has been a distinct tendency, in the past, to regard conservation of such areas as salt marsh as a process of maintenance of the *status quo*. Yet salt marshes, like all other coastal areas, are places of constant change (Doody in Jones, 1995): as we discussed on p. 107, salt marsh successions do not necessarily proceed continuously, but are interrupted and may be completely re-set to pioneer zones by such mechanisms as re-growth of drainage creeks or erosion of the salt marsh cliff. Conservation measures therefore need to incorporate an appreciation of the necessity for such change. In addition, if sea level does indeed rise by several millimetres each year as predicted, it will be virtually impossible to raise all the sea defences to cope with the problem. Instead, it may be more sensible to allow the sea to reclaim some areas of low-lying land at present protected by sea walls. Such 'managed retreat' would be relatively cheap, and would increase areas appropriate for nature conservation as low-lying pastures revert to salt marsh. This marsh, in turn, would act as a barrier against wave erosion without the need for 'hard' engineering structures.

The answers: treating, minimizing, and preventing damage

With such a plethora of abuses being hurled at the estuarine environment, it is not surprising that some estuaries have become little more than open sewers with very little living in them apart from bacteria and fungi. However, most man-made deterioration can be reversed to some degree, and we start by considering the Chesapeake Bay and the Thames as case histories.

Rehabilitation: the return of the Chesapeake Bay and the Thames

Deteriorating conditions in the Chesapeake Bay, characterized by areas of anoxia and reduced fish and shellfish harvests, triggered a programme of scientific investigation and restoration which began in 1976. The three areas of greatest concern were the release of toxic chemicals, nutrient enrichment (eutrophication), and the loss of submerged seagrass beds. Understanding the details of these problems has been difficult, particularly in the area of nutrient dynamics, where the relation between nitrogen input and anoxia remains unclear. A phosphate ban has been successful in reducing phosphate entering the Bay by 50%, but nitrogen levels have not yet been lowered. A moratorium on fishing for striped bass has helped local populations to recover, but oyster populations have not recovered, nor have seagrass beds begun to increase again (deFur, 1997). Partly the failures may be due to persistence of pathogenic organisms that have been introduced (Ruiz *et al.*, 1997), but evidently in the case of the Chesapeake Bay the reversal of degradation is a far from simple matter.

In the Thames estuary, UK, a long programme of reversing the impacts of humans has been to a great degree successful (Attrill, 1998). Deterioration of water quality in the nineteenth century was followed by the construction of sewers for London which carried the domestic and industrial effluents to outfalls downstream. As a result, fish returned to the Thames by the beginning of the twentieth century. But the population of London continued to rise, and standards of effluent treatment fell after this. With the advent of non-biodegradable synthetic detergents, the zone of oxygen depletion in the estuary increased to a record 42 km in the summer of 1949 (Tinsley in Attrill, 1998). From this time onwards, controlling authorities began to adopt a series of defined objectives in terms of dissolved oxygen, because lack of oxygen was seen as the major problem. Major expenditure on sewage works in the 1970s improved water quality. Storm-water discharges, which bring unregulated amounts of organic waste and toxins, were combated by introducing devices to aerate estuary water. More stringent controls for trade effluent were adopted. As a result of these actions, flounders and eels returned to the previously lifeless regions in 1964, and there has been a progressive increase in numbers of fish species since then (Fig. 11.8). In 1974, the first salmon for 140 years was recorded, and following stocking and rearing programmes, recorded salmon runs now vary from 50 to 300 each year (Thomas in Attrill, 1998).

The Thames is thus now able to support a healthy fish population. Partly this is due directly to a rise in dissolved oxygen concentrations. To a great degree, however, the growth of fish numbers reflects the effects of increased oxygen on benthic diversity, as benthic invertebrates provide a food source for the fish. There is now a much more diverse benthic fauna than 20 years ago (Attrill in Attrill, 1998). The Thames estuary has probably reached a steady state in terms of its rehabilitation, as long as the present measures controlling its water quality are maintained.

In conclusion, many of the influences of pollutants *can* be removed by building treatment plants for sewage effluents, reducing permissible limits for the emission of various toxic substances and attempting to control overall nutrient input. These processes take a long time, however, and we still lack basic understanding in some areas.

Fig. 11.8 Cumulative numbers of species recorded in the Thames estuary in a 55-km stretch (10 km above London Bridge to 45 km below it). Arrows show years in which flounders and salmon were first recorded. Many records were taken from the cooling water screens at power stations. Note that sampling effort increased in 1967, partly accounting for increased species numbers, but effort remained constant after this. (After Thomas in Attrill, 1998.)

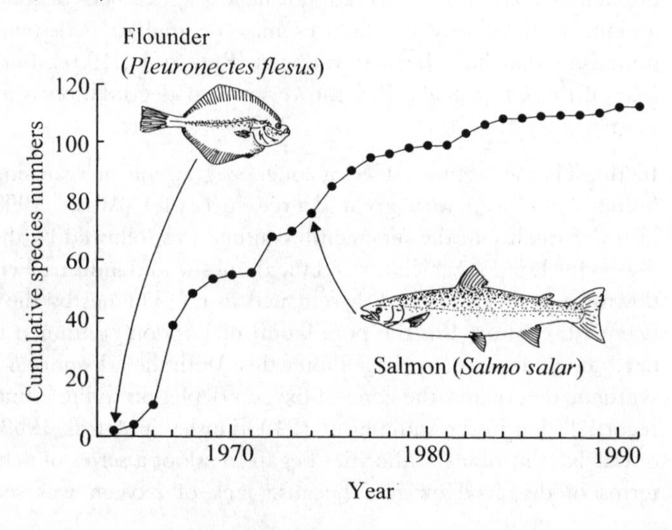

Conservation in estuaries

While it is important to reverse the effects of man-made degradation as far as possible, it is also important to protect those areas that have not so far been degraded. In many industrialized countries, it is now normal practice to commission environmental impact studies prior to any major development. For example, extension of harbour facilities, proposed barrages, or major land-claims are investigated with regard to the predicted effect they would have on the environment and the biota. Monitoring studies before and after the development should then follow the actual, as compared with the forecast, impact.

For sites that are already known to be important in terms of estuarine ecology, protection has usually been sought by creating some form of conservation area or 'marine protected area'. These vary from local nature reserves to those designated by international conventions, and present a bewildering diversity of legislation, accompanied by an ever-increasing array of acronyms (Jones *et al.*, 1996). Perhaps the most widely accepted internationally are the Ramsar sites (designated to protect areas of wetland, particularly in reference to waterfowl), and Biosphere reserves (promoted by UNESCO to protect a variety of environments). In Europe, the European Commission has created further reserves under its Habitats Directive and Birds Directive. In addition, many sites are designated under national laws. In Britain, for example, there are more than 300 'Sites of Special Scientific Interest' (SSSIs) associated with estuaries (Davidson *et al.*, 1991). There are also areas designated specifically as marine reserves, and these have considerable potential in safeguarding estuarine ecosystems because they link intertidal and subtidal areas.

There are very many other categories of sites that are 'protected'. But does any of this protection really work? Up to the present time, the answer, sadly, has to be 'no'. Protection under one law or another usually delays degradation, but in many cases permission to 'develop' is seen to be more important than the need to 'conserve'. Progressive nibbling at the edge of conservation sites may reduce their importance until they are no longer worth conserving. Areas such as SSSIs are frequently seen as wasteland that can be usefully reclaimed. For example, the Taff–Ely estuary, a subestuary of the Severn, contained an SSSI that provided roost sites and feeding areas for up to 6000 waterfowl (Tyler in Jones *et al.*, 1996). It has now been enclosed by a barrage which impounds water and covers the mudflats permanently. Although land elsewhere has been provided in compensation, this is unlikely to be an adequate replacement. In Portugal—and elsewhere—open sandy coastline and dunes have been eroded by erection of groynes that interrupt the supply of moving sand. Conservation areas are therefore diminished. Such inappropriate management raises the concept of 'sustainability': changes to coasts and estuaries should be constrained within a sustainable framework (Pethick in Jones *et al.*, 1996). This concept necessitates a change from the piecemeal approach so far discussed, and we go on now to consider ways in which management can be integrated.

An integrated approach: coastal zone management

The human population of coastal regions is estimated to double by the year 2020 (Pullen in Ormond *et al.*, 1992), so pressure on the interface between sea and land will rise inexorably. In the face of this pressure, what approach can be taken to protect biodiversity and the existence of the numerous habitats discussed in this book? The first step involves some kind of review of coastal habitats and the species they contain, so that areas of conservation interest can be identified. These may involve areas that have high diversity, or shelter rare species, as well as those containing typical examples of various habitats. For example, in the UK, the Nature Conservancy Council reviewed in great detail the knowledge of estuarine sites in Britain (Davidson *et al.*, 1991). Following such a review, sites can be graded or selected in terms of their biological interest, and protection considered accordingly.

Because different types of shore differ in their vulnerability to both natural and human impacts, account needs to be taken of how they respond to various kinds of change. For example, salt marshes are relatively robust in terms of their response to natural erosion (Pethick in Jones *et al.*, 1996), but are fragile in their response to significant oil pollution. Detailed monitoring programmes are needed to establish whether natural changes are taking place in such habitats, so that future changes can be assessed. In addition, to be prepared for catastrophic human impacts, the coast should be surveyed specifically with regard to

Fig. 11.9 Map showing designated priority areas in the Voordelta region of the Netherlands. The scheme extends outwards roughly to the 20 m depth contour, and includes coastal areas in which priority is given to recreation or to nature conservation. (After Jones in Jones *et al.*, 1996.)

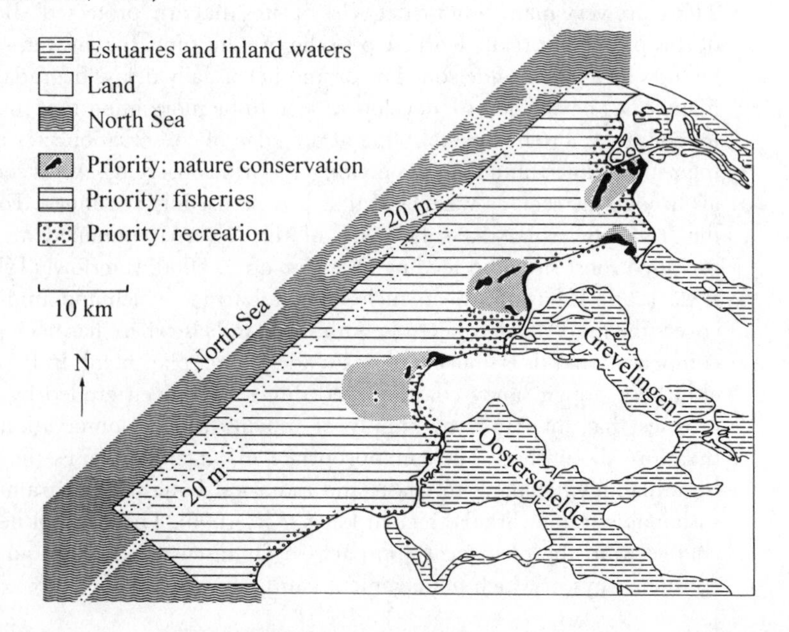

its vulnerability to, say, oil spills. For example, in Svalbard (an archipelago in the Arctic Ocean east of Greenland), a vulnerability index rates sheltered bays with abundant macrophytes very vulnerable, while exposed gravel beaches are much less so (Weslawski *et al.*, 1997). With the aid of such surveys, protective measures can be concentrated at the most vulnerable sites.

In any integrated scheme, these biological viewpoints must be coupled with reviews of available resources, and demand for use and for development. Provided *sustainable* development is proposed, it can then be incorporated in an integrated plan for management of a section of coastline. Such a plan must include an area of adjacent sea and an area of adjacent land, and forms what is known as Coastal Zone Management (CZM). Ideally, management should coordinate planning both locally and nationally, and this involves bringing together a wide variety of organizations into some kind of consortium (Baird, 1996).

In the Netherlands, a policy plan for the Voordelta region was agreed in 1993, involving the division into 'zones' of more than 80 km of coastline. The zones are defined as areas in which priority is given to one activity (Fig. 11.9). Thus some zones have an accent on nature conservation, others on recreation, while offshore regions have an accent on fisheries. The plan also involves the discussion of proposed developments and how these might be incorporated in a sustainable fashion. CZM in one form or another is now developing widely (Jones *et al.*, 1996), and seems to offer the only reasonable compromise between change and conservation.

Further reading

General

Coastal ecosystems and estuaries are considered in the general context of aquatic ecology by Barnes and Mann (1991), and in relation to oceanographic processes by Summerhayes and Thorpe (1996). A general account of intertidal ecology is provided by Raffaelli and Hawkins (1996). Ormond *et al.* (1997) discuss marine diversity, including soft sediments. Physical backgrounds to coastal soft sediments, estuaries, and shallow-water processes are given by Brown *et al.* (1997), Pethick (1984), and Trenhaile (1997). A very readable discussion of water flow near the sediment surface is provided by Snelgrove and Butman (1994).

Many journals cover the topic of soft sediments, but for estuaries the two most relevant are *Estuaries* and *Estuarine and coastal shelf science.* Current research interests are discussed by two major societies, the Estuarine Research Federation (USA) and the Estuarine and Coastal Sciences Association (based in the UK but with membership worldwide). Contact addresses for these societies can be found in the journals mentioned above.

Identification of flora and fauna

The list of identification guides given by Little and Kitching (1996) for rocky shores covers sources appropriate for soft shores. In addition, several regional accounts discuss and describe soft-shore biota and their ecology: temperate Australia, see Underwood and Chapman (1995); southern Africa, see Branch *et al.* (1994); Hong Kong, see Morton and Morton (1983); northeastern North America see Pollock (1998); the northern Pacific coast of North America, see Kozloff (1993); the brackish-water fauna of northwestern Europe is described by Barnes (1994*b*); life in the Chesapeake Bay, USA, is very readably described and copiously illustrated by Lippson and Lippson (1997).

Unvegetated sedimentary shores

The ecology of tidal mudflats is discussed in a very readable account by Reise (1985), and the ecology of marine deposit feeders is considered in relation to the environment provided by benthic sediments by Lopez *et al.* (1989). Sandy beach ecology is described by Brown and McLachlan (1990), and adaptations of some animal groups by Little (1990).

Salt marshes and mangrove swamps

The ecology of salt marsh flora and fauna is admirably discussed by Long and Mason (1983), who provide a very wide-ranging introduction. A more botanical account is given by Adam (1990). A variety of approaches including geomorphological background and conservation of salt marshes is provided by Allen and Pye (1992). The ecology of mangrove swamps is described in a very readable and complete account by Hogarth (1999), who discusses the importance of both animals and plants, and should be consulted for further detail. For a volume with many excellent colour photographs see Stafford-Deitsch (1996). The environments of mangrove swamps and salt marshes in relation to animal colonization are discussed by Little (1990).

Estuaries and lagoons

A physical introduction is provided by Dyer (1997). An account of estuarine ecology based on trophic levels is given by McLusky (1989), and Day *et al.* (1989) discuss the overall ecology of estuaries, mainly those in North America. Tropical estuaries and their fisheries are described by Blaber (1997). Although somewhat out of date, the volume by Green (1968) provides a wealth of information about estuarine animals. Barnes (1980) gives a short account of lagoonal biology. Kjerfve (1994) provides a number of articles about lagoonal processes.

Human impacts

The whole subject of marine pollution is excellently summarized by Clark (1997). Human effects upon estuarine shores, in the context of natural evolutionary processes, are discussed by Nordstrom and Roman (1996). The use of marine nature reserves or 'marine protected areas' is covered by Gubbay (1995).

References

Note that contributions to edited volumes are given in the text under the names of author(s) and editor(s) (e.g. Peterson in Livingston, 1979), but the reference appears in this list only under the name(s) of the editor(s).

Adam, P. (1990). *Saltmarsh ecology.* Cambridge University Press, Cambridge.

Albrecht, A.S. (1998). Soft bottom versus hard rock: community ecology of macroalgae on intertidal mussel beds in the Wadden Sea. *Journal of Experimental Marine Biology and Ecology,* **229**, 85–109.

Alldredge, A.L. and Silver, M.W. (1988). Characteristics, dynamics and significance of marine snow. *Progress in Oceanography,* **20**, 41–82.

Allen, D.M., Johnson, W.S., and Ogburn-Matthews, V. (1995). Trophic relationships and seasonal utilization of salt-marsh creeks by zooplanktivorous fishes. *Environmental Biology of Fishes*, **42**, 37–50.

Allen, J.R.L. and Pye, K. (ed.) (1992). *Saltmarshes: morphodynamics, conservation and engineering significance.* Cambridge University Press, Cambridge.

Allen, R.L. and Baltz, D.M. (1997). Distribution and microhabitat use by flatfishes in a Louisiana estuary. *Environmental Biology of Fishes*, **50**, 85–103.

Alongi, D.M. (1990). The ecology of tropical soft-bottom benthic ecosystems. *Oceanography and Marine Biology: An Annual Review,* **28**, 381–496.

Ambrose, W.G. (1984). Influences of predatory polychaetes and epibenthic predators on the structure of a soft-bottom community in a Maine estuary. *Journal of Experimental Marine Biology and Ecology,* **81**, 115–45.

Aneer, G. (1980). Estimates of feeding pressure on pelagic and benthic organisms by Baltic herring (*Clupea harengus* v. *membras* L.). *Ophelia*, Suppl. **1**, 265–275.

Anon. (1959). *Symposium on the classification of brackish waters.* Societas Internationalis Limnologiae. Venice.

Ansell, A.D. (1981). Functional morphology and feeding of *Donax serra* Röding and *Donax sordidus* Hanley (Bivalvia: Donacidae). *Journal of Molluscan Studies*, **47**, 59–72.

Attrill, M.J. (ed.) (1998). *A rehabilitated estuarine ecosystem. The environment and ecology of the Thames estuary.* Kluwer Academic Publishers, Dordrecht.

Bachelet, G., de Montaudouin, X., and Dauvin, J.-C. (1996). The quantitative distribution of subtidal macrozoobenthic assemblages in Arcachon Bay in relation to environmental factors: a multivariate analysis. *Estuarine and Coastal Shelf Science*, **42**, 371–391.

Bacon, G.S., MacDonald, B.A., and Ward, J.E. (1998). Physiological responses of infaunal (*Mya arenaria*) and epifaunal (*Placopecten magellanicus*) bivalves to variations in the concentration and quality of suspended particles. 1. Feeding activity and selection. *Journal of Experimental Marine Biology and Ecology,* **219**, 105–125.

Baird, D. and Milne, H. (1981). Energy flow in the Ythan estuary, Aberdeenshire, Scotland. *Estuarine and Coastal Shelf Science,* **13**, 455–472.

Baird, D. and Ulanowicz, R.E. (1989). The seasonal dynamics of the Chesapeake Bay ecosystem. *Ecological Monographs*, **59**, 329–364.

Baird, D., Evans, P.R., Milne, H., and Pienkowski, M.W. (1985). Utilisation by shorebirds of benthic invertebrate production in intertidal areas. *Oceanography and Marine Biology: An Annual Review*, **23**, 575–597.

Baird, R.C. (1996). Toward new paradigms in coastal resource management: linkages and institutional effectiveness. *Estuaries*, **19**, 320–335.

Baker, J.M. and Wolff, W.J. (ed.) (1987). *Biological surveys of estuaries and coasts.* Cambridge University Press, Cambridge.

Barnes, R.S.K. (ed.) (1977). *The coastline.* John Wiley and Sons, London.

Barnes, R.S.K. (1980). *Coastal lagoons.* Cambridge University Press, Cambridge.

Barnes, R.S.K. (1986). Daily activity rhythms in the intertidal gastropod *Hydrobia ulvae* (Pennant). *Estuarine and Coastal Shelf Science*, **22**, 325–334.

Barnes, R.S.K. (1988). The faunas of land-locked lagoons: chance differences and the problems of dispersal. *Estuarine and Coastal Shelf Science*, **26**, 309–318.

Barnes, R.S.K. (1989). What, if anything, is a brackish-water fauna? *Transactions of the Royal Society of Edinburgh, Earth Sciences*, **80**, 235–240.

Barnes, R.S.K. (1994*a*). Investment in eggs of lagoonal *Hydrobia ventrosa* and life-history strategies of north-west European *Hydrobia* species. *Journal of the Marine Biological Association of the UK*, **74**, 637–650.

Barnes, R.S.K. (1994*b*). *The brackish-water fauna of northwestern Europe.* Cambridge University Press, Cambridge.

Barnes, R.S.K. and Greenwood, J.G. (1978). The response of the intertidal gastropod *Hydrobia ulvae* (Pennant) to sediments of differing particle size. *Journal of Experimental Marine Biology and Ecology*, **31**, 43–54.

Barnes, R.S.K. and Mann, K.H. (ed.) (1991). *Fundamentals of aquatic ecology.* Blackwell, Oxford.

Berry, A.J. (1964). Faunal zonation in mangrove swamps. *Bulletin of the National Museum of Singapore*, **32**, 90–98.

Bertness, M.D. (1985). Fiddler crab regulation of *Spartina alterniflora* production on a New England saltmarsh. *Ecology*, **66**, 1042–1055.

Blaber, S.J.M. (1997). *Fish and fisheries of tropical estuaries.* Chapman and Hall, London.

Blaber, S.J.M., Farmer, M.J., Milton, D.A., Pang, J., Boon-Teck, O., and Wong, P. (1997). The ichthyoplankton of selected estuaries in Sarawak and Sabah: composition, distribution and habitat affinities. *Estuarine and Coastal Shelf Science*, **45**, 197–208.

Black, K.S. and Paterson, D.M. (1997). Measurement of the erosion potential of cohesive marine sediments: a review of current *in situ* technology. *Journal of Marine Environmental Engineering*, **4**, 43–83.

Blandford, P. and Little, C. (1983). Salinity detection by *Hydrobia ulvae* (Pennant) and *Potamopyrgus jenkinsi* Smith (Gastropoda: Prosobranchia). *Journal of Experimental Marine Biology and Ecology*, **68**, 25–38.

Bonsdorff, E. and Blomqvist, E.M. (1993). Biotic couplings on shallow water soft bottoms—examples from the northern Baltic Sea. *Oceanography and Marine Biology: An Annual Review*, **31**, 153–176.

Bonsdorff, E., Blomqvist, E.M., Mattila, J., and Norkko, A. (1997). Coastal eutrophication: causes, consequences and perspectives in the archipelago areas of the northern Baltic Sea. *Estuarine and Coastal Shelf Science*, **44** (Suppl. A), 63–72.

Boyden, C.R. (1972). The behaviour, survival and respiration of the cockles *Cerastoderma edule* and *C. glaucum* in air. *Journal of the Marine Biological Association of the UK*, **52**, 661–680.

Boyden, C.R. and Little, C. (1973). Faunal distributions in soft sediments of the Severn estuary. *Estuarine and Coastal Marine Science*, **1**, 203–223.

Branch, G.M., Griffiths, C.L., Branch, M.L., and Beckley, L.E. (1994). *Two oceans. A guide to the marine life of southern Africa.* David Philip, Cape Town.

Bregazzi, P.K. and Naylor, E. (1972). The locomotor activity rhythm of *Talitrus saltator* (Montagu) (Crustacea, Amphipoda). *Journal of Experimental Biology*, **57**, 375–391.

Brown, A.C. and McLachlan, A. (1990). *Ecology of sandy shores.* Elsevier, Amsterdam.

Brown, A.C. and Trueman, E.R. (1991). Burrowing of sandy-beach molluscs in relation to penetrability of the substratum. *Journal of Molluscan Studies*, **57**, 134–136.

Brown, A.C., Stenton-Dozey, J.M.E., and Trueman, E.R. (1989). Sandy-beach bivalves and gastropods: a comparison between *Donax serra* and *Bullia digitalis*. *Advances in Marine Biology*, **25**, 179–247.

Brown, B. and Wilson, W.H. (1997). The role of commercial digging of mudflats as an agent for change of infaunal intertidal populations. *Journal of Experimental Marine Biology and Ecology*, **218**, 49–61.

Brown, J., Colling, A., Park, D., Phillips, J., Rothery, D., and Wright, J. (1997). *Waves, tides and shallow-water processes.* Butterworth-Heinemann, Oxford and The Open University, Milton Keynes.

Bulger, A.J., Hayden, B.P., Monaco, M.E., Nelson, D.M., and McCormick-Ray, M.G. (1993). Biologically-based estuarine salinity zones derived from a multivariate analysis. *Estuaries*, **16**, 311–322.

Bulnheim, H.-P. (1991). Zur Ökophysiologie, Sexualität und Populationsgenetik litoraler Gammaridea—ein Überblick. *Helgolander Meeresuntersuchungen*, **45**, 381–401.

Butler, H. and Rogerson, A. (1997). Consumption rates of six species of marine benthic naked amoebae (Gymnamoebia) from sediments in the Clyde Sea area. *Journal of the Marine Biological Association of the UK*, **77**, 989–997.

Chen, Y-H., Shaw, P-T., and Wolcott, T.G. (1997). Enhancing estuarine retention of planktonic larvae by tidal currents. *Estuarine and Coastal Shelf Science*, **45**, 525–533.

Cherrill, A.J. and James, R. (1987). Evidence for competition between mudsnails (Hydrobiidae): a field experiment. *Hydrobiologia*, **150**, 25–31.

Claridge, P.N., Potter, I.C., and Hardisty, M.W. (1986). Seasonal changes in movements, abundance, size composition and diversity of the fish fauna of the Severn estuary. *Journal of the Marine Biological Association of the UK*, **66**, 229–258.

Clark, R.B. (1997). *Marine pollution.* 4th edn. Clarendon Press, Oxford.

Collins, N.R. and Williams, R. (1981). Zooplankton of the Bristol Channel and Severn estuary. The distribution of four copepods in relation to salinity. *Marine Biology*, **64**, 273–283.

Coull, B.C. (ed.) (1977). *Ecology of marine benthos.* University of South Carolina Press, Columbia, S.C.

Crawford, R.M., Dorey, A.E., Little, C., and Barnes, R.S.K. (1979). Ecology of Swanpool, Falmouth V. Phytoplankton and nutrients. *Estuarine and Coastal Marine Science*, **9**, 135–160.

Créach, V., Schricke, M.T., Bertru, G., and Mariotti, A. (1997). Stable isotopes and gut analyses to determine feeding relationships in saltmarsh macroconsumers. *Estuarine and Coastal Shelf Science*, **44**, 599–611.

Cummings, V.J., Pridmore, R.D., Thrush, S.F., and Hewitt, J.E. (1996). Effect of the spionid polychaete *Boccardia syrtis* on the distribution and survival of juvenile *Macomona liliana* (Bivalvia: Tellinacea). *Marine Biology*, **126**, 91–98.

Curtis, L.A. and Hubbard, K.M.K. (1993). Species relationships in a marine gastropod-trematode ecological system. *Biological Bulletin of the Marine Biological Laboratory, Woods Hole*, **184**, 25–35.

Daborn, G.R., Amos, C.L., Brylinsky, M., Christian, H., Drapeau, G., Faas, R.W., Grant, J., Long, B., Paterson, D.M., and Perillo, G.M.E. (1993). An ecological cascade effect: migratory birds affect stability of intertidal sediments. *Limnology and Oceanography*, **38**, 225–231.

Dadswell, M.J. and Rulifson, R.A. (1994). Macrotidal estuaries: a region of collision between migratory marine animals and tidal power development. *Biological Journal of the Linnean Society*, **51**, 93–113.

Dauer, D.M. (1985). Functional morphology and feeding behavior of *Paraprionospio pinnata* (Polychaeta: Spionidae). *Marine Biology*, **85**, 143–151.

Davenport, J., Gruffydd, Ll.D., and Beaumont, A.R. (1975). An apparatus to supply water of fluctuating salinity and its use in a study of the salinity tolerances of larvae of the scallop *Pecten maximus* L. *Journal of the Marine Biological Association of the UK*, **55**, 391–409.

Davidson, N.C., Laffoley, D.d'A., Doody, J.P., Way, L.S., Gordon, J., Key, R., Drake, C.M., Pienkowski, M.W., Mitchell, R., and Duff, K.L. (1991). *Nature conservation and estuaries in Great Britain*. Nature Conservancy Council, Peterborough.

Day, J.W., Hall, C.A.S., Kemp, W.M., and Yanez-Arancibia, A. (ed.) (1979). *Estuarine ecology*. Wiley, New York.

Defeo, O., Brazeiro, A., de Alava, A., and Riestra, G. (1997). Is sandy beach macrofauna only physically controlled? Role of substrate and competition in isopods. *Estuarine and Coastal Shelf Science*, **45**, 453–462.

deFur, P.L. (1997). The Chesapeake Bay program: an example of ecological risk assessment. *American Zoologist*, **37**, 641–649.

Desroy, N., Retière, C., and Thiébaut, E. (1998). Infaunal predation regulates benthic recruitment: an experimental study of the influence of the predator *Nephtys hombergii* (Savigny) on recruits of *Nereis diversicolor* (O.F.Müller). *Journal of Experimental Marine Biology and Ecology*, **228**, 257–272.

Diaz, R.J. and Rosenberg, R. (1995). Marine benthic hypoxia: a review of its ecological effects and the behavioural responses of benthic macrofauna. *Oceanography and Marine Biology: An Annual Review*, **33**, 245–303.

Dick, J.T.A., Irvine, D.E., and Elwood, R.W. (1990). Differential predation by males on moulted females may explain the competitive displacement of *Gammarus duebeni* by *G. pulex* (Amphipoda). *Behavioural Ecology and Sociobiology*, **26**, 41–45.

Dring, M.J. (1982). *The biology of marine plants*. Arnold, London.

Dyer, K.R. (ed.) (1979). *Estuarine hydrography and sedimentation*. Cambridge University Press, Cambridge.

Dyer, K.R. (1997). *Estuaries—a physical introduction*. 2nd edn. John Wiley and Sons, Chichester.

Eckman, J.E. (1985). Flow disruption by animal-tube mimic affects sediment bacterial colonization. *Journal of Marine Research*, **43**, 419–435.

Edelvang, K. and Austen, I. (1997). The temporal variation of flocs and fecal pellets in a tidal channel. *Estuarine and Coastal Shelf Science*, **44**, 361–367.

Edwards, R.R.C. (1978). Ecology of a coastal lagoon complex in Mexico. *Estuarine and Coastal Marine Science*, **6**, 75–92.

Elliott, M. and Kingston, P.F. (1987). The sublittoral benthic fauna of the estuary and Firth of Forth, Scotland. *Proceedings of the Royal Society of Edinburgh*, **93B**, 449–465.

Elliott, M., O'Reilly, M.G., and Taylor, C.J.L. (1990). The Forth estuary: a nursery and overwintering area for the North Sea fishes. *Hydrobiologia*, **195**, 89–103.

El-Sabh, M.I., Aung, T.H., and Murty, T.S. (1997). Physical processes in inverse estuarine systems. *Oceanography and Marine Biology: An Annual Review*, **35**, 1–69.

Fenchel, T. (1969). The ecology of marine microbenthos. IV. Structure and function of the benthic ecosystem, its physical factors and the microfauna communities with special reference to the ciliated Protozoa. *Ophelia*, **6**, 1–182.

Fenchel, T. (1970). Studies on the decomposition of organic detritus derived from the turtle grass *Thalassia testudinum*. *Limnology and Oceanography*, **15**, 14–20.

Fenchel, T. (1975). Character displacement and coexistence in mud snails (Hydrobiidae). *Oecologia (Berlin)*, **20**, 19–32.

Fenchel, T. and Finlay, B.J. (1995). *Ecology and evolution in anoxic worlds*. Oxford University Press, Oxford.

Fenchel, T.M. and Jørgensen, B.B. (1977). Detrital food chains of aquatic ecosystems: the role of bacteria. *Advances in Microbial Ecology*, **1**, 1–58.

Ferns, P.N. (1992). *Bird life of coasts and estuaries*. Cambridge University Press, Cambridge.

Flemer, D.A., Livingston, R.J., and McGlynn, S.E. (1998). Seasonal growth stimulation of sub-temperate estuarine phytoplankton to nitrogen and phosphorus: an outdoor microcosm experiment. *Estuaries*, **21**, 145–159.

Flowers, T.J., Troke, P.F., and Yeo, A.R. (1977). The mechanism of salt tolerance in halophytes. *Annual Review of Plant Physiology*, **28**, 89–121.

Fonseca, M.S. (1989). Sediment stabilization by *Halophila decipiens* in comparison to other sea grasses. *Estuarine and Coastal Shelf Science*, **29**, 501–507.

Fromentin, J.M., Ibanez, F., Dauvin, J.C., Dewarumez, J.M., and Elkaim, B. (1997). Long-term changes of four macrobenthic assemblages from 1978 to 1992. *Journal of the Marine Biological Association of the UK*, **77**, 287–310.

Furukawa, F., Wolanski, E., and Mueller, H. (1997). Currents and sediment transport in mangrove forests. *Estuarine and Coastal Shelf Science*, **44**, 301–310.

Gamble, F.W. and Keeble, F. (1903). The bionomics of *Convoluta roscoffensis*, with special reference to its green cells. *Quarterly Journal of Microscopical Science*, **47**, 363–431.

Gattuso, J.-P., Frankignoulle, M., and Wollast, R. (1998). Carbon and carbonate metabolism in coastal aquatic ecosystems. *Annual Review of Ecology and Systematics*, **29**, 405–434.

Giere, O. (1993). *Meiobenthology: the microscopic fauna in aquatic science*. Springer-Verlag, Berlin.

Gillibrand, P.A., Turrell, W.R., Moore, D.C., and Adams, R.D. (1996). Bottom water stagnation and oxygen depletion in a Scottish sea loch. *Estuarine and Coastal Shelf Science*, **43**, 217–235.

Gorbushin, A.M. (1997). Field evidence of trematode-induced gigantism in *Hydrobia* spp. (Gastropoda: Prosobranchia). *Journal of the Marine Biological Association of the UK*, **77**, 785–800.

Gordon, D.C. (1994). Intertidal ecology and potential power impacts, Bay of Fundy, Canada. *Biological Journal of the Linnean Society*, **51**, 17–23.

Grahame, J. (1987). *Plankton and fisheries*. Arnold, London.

Grall, J. and Glémarec, M. (1997). Using biotic indices to estimate macrobenthic community perturbations in the Bay of Brest. *Estuarine and Coastal Shelf Science*, **44** (Suppl. A), 43–53.

Grant, A. and Briggs, A.D. (1998). Use of ivermectin in marine fish farms: some concerns. *Marine Pollution Bulletin*, **36**, 566–568.

Grant, J., Cranford, P., and Emerson, C. (1997). Sediment resuspension rates, organic matter quality and food utilization by sea scallops (*Placopecten magellanicus*) on Georges Bank. *Journal of Marine Research*, **55**, 965–994.

Gray, A. (ed.) (1992). *The ecological impact of estuarine barrages.* British Ecological Society and Field Studies Council, Montford Bridge, Shrewsbury.

Gray, A.J., Marshall, D.F., and Raybould, A.F. (1991). A century of evolution in *Spartina anglica. Advances in Ecological Research*, **21**, 1–62.

Gray, J.S. (1981). *The ecology of marine sediments.* Cambridge University Press, Cambridge.

Gray, J.S. (1992). Eutrophication in the sea. In *Marine eutrophication and population dynamics* (ed. G. Colombo, I. Ferrari, V.U. Ceccerelli, and R. Rossi), pp. 3–15. Olsen and Olsen, Fredensborg.

Green, J. (1968). *The biology of estuarine animals.* Sidgwick and Jackson, London.

Gubbay, S. (ed.) (1995). *Marine protected areas.* Chapman and Hall, London.

Hale, W.G. (1980). *Waders.* Collins, London.

Hall, S.J. (1994). Physical disturbance and marine benthic communities: life in unconsolidated sediments. *Oceanography and Marine Biology: An Annual Review*, **32**, 179–239.

Hall, S.J. and Raffaelli, D.G. (1991). Food web patterns: lessons from a species rich web. *Journal of Animal Ecology*, **60**, 823–841.

Hall, S.J., Robertson, M.R., Basford, D.J., and Fryer, R. (1993). Pit digging by the crab *Cancer pagurus*: a test for long-term, large-scale effects on infaunal community structure. *Journal of Animal Ecology*, **62**, 59–66.

Hall, S.J., Raffaelli, D., and Thrush, S.F. (1994). Patchiness and disturbance in shallow water benthic assemblages. In *Aquatic ecology. Scale, pattern and process* (ed. P.S. Giller, A.G. Hildrew and D.G. Raffaelli), pp. 333–375. Blackwell, Oxford.

Hansen, K. and Kristensen, E. (1997). Impact of macrofaunal recolonization on benthic metabolism and nutrient fluxes in a shallow marine sediment previously overgrown with macroalgal mats. *Estuarine and Coastal Shelf Science*, **45**, 613–628.

Hartwick, R.F. (1976). Beach orientation in talitrid amphipods: capacities and strategies. *Behavioural Ecology and Sociobiology*, **1**, 447–458.

Harvey, H.W. (1963). *The chemistry and fertility of sea waters.* Cambridge University Press, Cambridge.

Hawkins, A.B. and Sebbage, M.J. (1972). The reversal of sand waves in the Bristol Channel. *Marine Geology*, **12**, M7–M9.

Hayes, M.O. (1975). Morphology of sand accumulation in estuaries: an introduction to the symposium. In *Estuarine research* (ed. L.E. Cronin), Vol. ii pp. 3–22. Academic Press, New York.

Head, P.C. (ed.) (1985). *Practical estuarine chemistry.* Cambridge University Press, Cambridge.

Heip, C.H.R., Goosen, N.K., Herman, P.M.J., Kromkamp, J., Middelburg, J.J., and Soetaert, K. (1995). Production and consumption of biological particles in temperate tidal estuaries. *Oceanography and Marine Biology: An Annual Review*, **33**, 1–149.

Henderson, P.A. and Holmes, R.H.A. (1989). Whiting migration in the Bristol Channel: a predator-prey relationship. *Journal of Fish Biology*, **34**, 409–416.

Henderson, P.A., James, D., and Holmes, R.H.A. (1992). Trophic structure within the Bristol Channel: seasonality and stability in Bridgwater Bay. *Journal of the Marine Biological Association of the UK*, **72**, 675–690.

Heymans, J.J. and Baird, D. (1995). Energy flow in the Kromme estuarine ecosystem, St Francis Bay, South Africa. *Estuarine and Coastal Shelf Science*, **41**, 39–59.

Heymans, J.J. and McLachlan, A. (1996). Carbon budget and network analysis of a high-energy beach/surf-zone ecosystem. *Estuarine and Coastal Shelf Science*, **43**, 485–505.

Hogarth, P.J. (1999). *Biology of mangroves.* Oxford University Press, Oxford.

Holligan, P.M. and Reiners, W.A. (1992). Predicting the responses of the coastal zone to global change. *Advances in Ecological Research*, **22**, 211–255.

Holme, N.A. and McIntyre, A.D. (ed.) (1984). *Methods for the study of marine benthos* (2nd edn). Blackwell, Oxford.

Holmström, W.F. and Morgan, E. (1983). Variation in the naturally occurring rhythm of the estuarine amphipod, *Corophium volutator* (Pallas). *Journal of the Marine Biological Association of the UK*, **63**, 833–850.

Holst, H., Zimmerman, H., Kausch, H., and Koste, W. (1998). Temporal and spatial dynamics of planktonic rotifers in the Elbe estuary during spring. *Estuarine and Coastal Shelf Science*, **47**, 261–273.

Hough, A.R. and Naylor, E. (1991). Field studies on retention of the planktonic copepod *Eurytemora affinis* in a mixed estuary. *Marine Ecology Progress Series*, **76**, 115–122.

Hughes, R.G. and Gerdol, V. (1997). Factors affecting the distribution of the amphipod *Corophium volutator* in two estuaries in south-east England. *Estuarine and Coastal Shelf Science*, **44**, 621–627.

Hughes, R.N. (1969). A study of feeding in *Scrobicularia plana*. *Journal of the Marine Biological Association of the UK*, **49**, 805–823.

Hulberg, L.W. and Oliver, J.S. (1980). Caging manipulations in marine soft-bottom communities: importance of animal interactions or sedimentary habitat modification. *Canadian Journal of Aquatic Sciences*, **37**, 1130–1139.

Humborg, G.C. (1997). Primary productivity regime and nutrient removal in the Danube estuary. *Estuarine and Coastal Shelf Science*, **45**, 579–589.

Hutchings, P.A. and Recher, H.F. (1982). The fauna of Australian mangroves. *Proceedings of the Linnean Society of New South Wales*, **106**, 83–121.

Hylleberg, J. (1975). Selective feeding by *Abarenicola pacifica* with notes on *Abarenicola vagabunda* and a concept of gardening in lugworms. *Ophelia*, **14**, 113–137.

James, R.J. and Fairweather, P.G. (1996). Spatial variation of intertidal macrofauna on a sandy ocean beach in Australia. *Estuarine and Coastal Shelf Science*, **43**, 81–107.

Jefferies, R.L. and Davy, A.J. (ed.) (1979). *Ecological processes in coastal environments*. Blackwell, Oxford.

Jenkins, G.P. and Wheatley, M.J. (1998). The influence of habitat structure on nearshore fish assemblages in a southern Australian embayment: comparison of shallow seagrass, reef-algal and unvegetated sand habitats, with emphasis on their importance to recruitment. *Journal of Experimental Marine Biology and Ecology*, **221**, 147–172.

Jensen, K.T. and Siegismund, H.R. (1980). The importance of diatoms and bacteria in the diet of *Hydrobia*-species. *Ophelia*, Suppl. **1**, 193–199.

Joint, I.R. (1984). The microbial ecology of the Bristol Channel. *Marine Pollution Bulletin*, **15**, 62–66.

Jones, N.V. (ed.) (1995). *The changing coastline*. (Coastal Zone Topics: Process, Ecology and Management **1**.) Joint Nature Conservation Committee and Estuarine and Coastal Sciences Association, Peterborough.

Jones, N.V. and Wolff, W.J. (ed.) (1981). *Feeding and survival strategies of estuarine organisms*. Plenum Press, New York.

Jones, P.S., Healy, M.G., and Williams, A.T. (ed.) (1996). *Studies in European coastal management*. Samara Publishing Ltd, Cardigan.

Jones, S.E. and Jago, C.F. (1993). In situ assessment of modification of sediment properties by burrowing invertebrates. *Marine Biology*, **115**, 133–142.

Jørgensen, B.B. and Des Marais, D.J. (1986). A simple fiber-optic microprobe for high resolution light measurements: application in marine sediment. *Limnology and Oceanography*, **31**, 1376–1383.

Khfaji, A.K. and Norton, T.A. (1979). The effects of salinity on the distribution of *Fucus ceranoides*. *Estuarine and Coastal Marine Science*, **8**, 433–439.

Kinne, O. (1971). Salinity. Animals—invertebrates. In *Marine ecology* (ed. O. Kinne), Vol. I, pp. 821–995.

Kirby, R. (1988). High concentration suspension (fluid mud) layers in estuaries. In *Physical processes in estuaries* (ed. J. Dronkers and W. Van Leussen), pp. 463–487. Springer-Verlag, Berlin.

Kirby, R. (1994). The evolution of the fine sediment regime of the Severn estuary and Bristol Channel. *Biological Journal of the Linnean Society*, **51**, 37–44.

Kirby, R. and Parker, W.R. (1983). Distribution and behavior of fine sediment in the Severn estuary and inner Bristol Channel, U.K. *Canadian Journal of Fisheries and Aquatic Sciences*, **40**, suppl. 1, 83–95.

Kitching, J.A., Ebling, F.J., Gamble, J.C., Hoare, R., McLeod, A.A.Q.R., and Norton, T.A. (1976). The ecology of Lough Ine XIX. Seasonal changes in the western trough. *Journal of Animal Ecology*, **45**, 731–758.

Kjerfve, B. (ed.) (1994). *Coastal lagoon processes.* Elsevier, Amsterdam.

Kneib, R.T. (1991). Indirect effects in experimental studies of marine soft-sediment communities. *American Zoologist*, **31**, 874–885.

Knights, B. (1979). Reclamation in the Netherlands. In *Estuarine and coastal land reclamation and water storage* (ed. B. Knights and A.J. Phillips), pp. 209–233. Saxon House, Farnborough.

Kostaschuk, R.A., Church, M.A., and Luternauer, J.L. (1992). Sediment transport over salt-wedge intrusions: Fraser River estuary, Canada. *Sedimentology*, **39**, 305–317.

Kozloff, E.N. (1993). *Seashore life of the northern Pacific coast.* University of Washington Press, Seattle.

Kramer, K.J.M., Brockman, U.H., and Warwick, R.M. (1994). *Tidal estuaries: manual of sampling and analytical procedures.* A.A. Balkema, Rotterdam.

Kristensen, E. (1988). Factors influencing the distribution of nereid polychaetes in Danish coastal waters. *Ophelia*, **29**, 127–140.

Lawrie, S.M. and Raffaelli, D.G. (1998). In situ swimming behaviour of the amphipod *Corophium volutator* (Pallas). *Journal of Experimental Marine Biology and Ecology*, **224**, 237–251.

Lee, S.Y. (1998). Ecological role of grapsid crabs in mangrove ecosystems: a review. *Marine and Freshwater Research*, **49**, 335–343.

Lennon, G.W., Bowers, D.G., Nunes, R.A., Scott, B.D., Ali, M., Boyle, J., Wenju, C., Herzfeld, M., Johansson, G., Nield, S., Petrusevics, P., Stephenson, P., Suskin, A.A., and Wijffels, S.E.A. (1987). Gravity currents and the release of salt from an inverse estuary. *Nature (London)*, **327**, 695–697.

Levin, L., Blair, N., DeMaster, D., Plaia, G., Fornes, W., Martin, C., and Thomas, C. (1997). Rapid subduction of organic matter by maldanid polychaetes on the North Carolina slope. *Journal of Marine Research*, **55**, 595–611.

Lippson, A.J. and Lippson, R.L. (1997). *Life in the Chesapeake Bay.* The Johns Hopkins University Press, Baltimore.

Little, C. (1984). Ecophysiology of *Nais elinguis* (Oligochaeta) in a brackish-water lagoon. *Estuarine and Coastal Shelf Science*, **18**, 231–244.

Little, C. (1986). Fluctuations in the meiofauna of the Aufwuchs community in a brackish-water lagoon. *Estuarine and Coastal Shelf Science*, **23**, 263–276.

Little, C. (1990). *The terrestrial invasion. An ecophysiological approach to the origins of land animals.* Cambridge University Press, Cambridge.

Little, C. (1992). Animals of Severn estuary salt marshes. *Proceedings of the Bristol Naturalists' Society* (1990), **50**, 83–94.

Little, C. and Kitching, J.A. (1996). *The biology of rocky shores.* Oxford University Press, Oxford.

Little, C. and Smith, L.P. (1980). Vertical zonation on rocky shores in the Severn estuary. *Estuarine and Coastal Marine Science*, **11**, 651–669.

Little, D.I. and Smith, J. (1994). Appraisal of contaminants in sediments of the inner Bristol Channel and Severn estuary. *Biological Journal of the Linnean Society*, **51**, 55–69.

Livingston, R.J. (ed.) (1979). *Ecological processes in coastal and marine systems.* Plenum Press, New York.

Long, S.P. and Mason, C.F. (1983). *Saltmarsh ecology.* Blackie, Glasgow.

Lopez, G.R. and Kofoed, L.H. (1980). Epipsammic browsing and deposit-feeding in mud snails (Hydrobiidae). *Journal of Marine Research*, **38**, 585–599.

Lopez, G., Taghon, G., and Levinton, J. (ed.) (1989). *Ecology of marine deposit feeders.* Springer-Verlag, Berlin.

Lüttge, U. (1975). Salt glands. In *Ion transport in plant cells and tissues* (ed. D.A. Baker and J.L. Hall), pp. 335–376. North-Holland Publishing Co., Amsterdam.

MacGinitie, G.E. and MacGinitie, N. (1949). *Natural history of marine animals.* McGraw-Hill, New York.

McLachlan, A. and Jaramillo, E. (1995). Zonation on sandy beaches. *Oceanography and Marine Biology: An Annual Review*, **33**, 305–335.

McLachlan, A., Wooldridge, T., and Van der Horst, G. (1979). Tidal movements of the macrofauna on an exposed sandy beach in South Africa. *Journal of Zoology (London)*, **187**, 433–442.

McLusky, D.S. (1989). *The estuarine ecosystem* (2nd edn). Blackie, Glasgow.

McLusky, D.S., de Jonge, V.N., and Pomfret, J. (ed.) (1990). *North Sea-estuaries interactions.* Kluwer Academic Publishers, Dordrecht.

McManus, J. and Elliot, M. (ed.) (1989). *Developments in estuarine and coastal study techniques.* Olsen and Olsen, Fredensborg.

Macnae, W. and Kalk, M. (1962). The ecology of mangrove swamps at Inhaca Island, Moçambique. *Journal of Ecology*, **50**, 19–34.

Maes, J., Taillieu, A., Van Damme, P.A., Cottenie, K., and Ollevier, F. (1998). Seasonal patterns in the fish and crustacean community of a turbid temperate estuary (Zeeschelde estuary, Belgium). *Estuarine and Coastal Shelf Science*, **47**, 143–151.

Maitland, P.S. and Campbell, R.N. (1992). *Freshwater fishes of the British Isles.* Harper Collins, London.

Malone, T.C., Conley, D.J., Fisher, T.R., Glibert, P.M., Harding, L.W., and Sellner, K.G. (1996). Scales of nutrient-limited phytoplankton productivity in Chesapeake Bay. *Estuaries*, **19**, 371–385.

Manahan, D.T. (1990). Adaptations by invertebrate larvae for nutrient acquisition from seawater. *American Zoologist*, **30**, 147–160.

Mann, K.H. (1988). Production and use of detritus in various freshwater, estuarine, and coastal marine ecosystems. *Limnology and Oceanography*, **33**, 910–930.

Marshall, S. and Elliott, M. (1998). Environmental influences on the fish assemblage of the Humber estuary, U.K. *Estuarine and Coastal Shelf Science*, **46**, 175–184.

Mayer, L.M., Schick, L.L., and Setchell, F.W. (1986). Measurement of protein in near-shore sediments. *Marine Ecology Progress Series*, **30**, 159–165.

Meadows, A., Meadows, P.S., Muir Wood, D., and Murray, J.M.H. (1994). Microbial effects on slope stability: an experimental analysis. *Sedimentology*, **41**, 423–435.

Meadows, P.S. and Reid, A. (1966). The behaviour of *Corophium volutator* (Crustacea, Amphipoda). *Journal of Zoology (London)*, **150**, 387–399.

Mettam, C. (1980). The intertidal ecosystem. In *An environmental appraisal of tidal power stations* (ed. T.L. Shaw), pp. 87–108. Pitman, London.

Meyers, M.B., Fossing, H., and Powell, E.N. (1987). Microdistribution of interstitial meiofauna, oxygen and sulfide gradients, and the tubes of macro-infauna. *Marine Ecology Progress Series*, **35**, 223–241.

Mille, G., Munoz, D., Jacquot, F., Rivet, L., and Bertrand, J.-C. (1998). The Amoco-Cadiz oil spill: evolution of petroleum hydrocarbons in the Ile Grande salt marshes (Brittany) after a 13–year period. *Estuarine and Coastal Shelf Science*, **47**, 547–559.

Monbet, Y. (1992). Control of phytoplankton biomass in estuaries: a comparative analysis of microtidal and macrotidal estuaries. *Estuaries*, **15**, 563–571.

Moreira, F. (1997). The importance of shorebirds to energy fluxes in a food web of a South European estuary. *Estuarine and Coastal Shelf Science*, **44**, 67–78.

Morrisey, D.J. (1987). Effect of population density and presence of a potential competitor on the growth rate of the mud snail *Hydrobia ulvae* (Pennant). *Journal of Experimental Marine Biology and Ecology*, **108**, 275–295.

Morton, B. and Morton, J. (1983). *The sea shore ecology of Hong Kong.* Hong Kong University Press, Hong Kong.

Newell, R.C. (1979). *Biology of intertidal animals* (3rd edn). Marine Ecological Surveys Ltd, Kent.

Newell, R.C., Seiderer, L.J., and Hitchcock, D.R. (1998). The impact of dredging works in coastal waters: a review of the sensitivity to disturbance and subsequent recovery of biological resources on the sea bed. *Oceanography and Marine Biology: An Annual Review*, **36**, 127–178.

Nichols, F.H., Thompson, J.K., and Shemel, L.E. (1990). Remarkable invasion of San Francisco Bay (California, USA) by the asian clam *Potamocorbula amurensis*. II. Displacement of a former community. *Marine Ecology Progress Series*, **66**, 95–101.

Nordstrom, K.F. and Roman, C.T. (ed.) (1996). *Estuarine shores. Evolution, environments and human alterations.* Wiley, Chichester.

O'Donohue, M.J.H. and Dennison, W.C. (1997). Phytoplankton productivity response to nutrient concentrations, light availability and temperature along an Australian estuarine gradient. *Estuaries*, **20**, 521–533.

Ogawa, N. and Ogura, N. (1997). Dynamics of particulate organic matter in the Tamagawa estuary and inner Tokyo Bay. *Estuarine and Coastal Shelf Science*, **44**, 263–273.

Okamura, B. (1990). Behavioural plasticity in the suspension feeding of benthic animals. In *Behavioural mechanisms of food selection* (ed. R.N. Hughes) pp 637–660. Springer-Verlag, Berlin.

Olafsson, E.B. and Persson, L.-E. (1986). The interaction between *Nereis diversicolor* O.F. Müller and *Corophium volutator* Pallas as a structuring force in a shallow brackish sediment. *Journal of Experimental Marine Biology and Ecology*, **103**, 103–117.

Olafsson, E.B., Peterson, C.H., and Ambrose, W.G. (1994). Does recruitment limitation structure populations and communities of macro-invertebrates in marine soft-sediments: the relative significance of pre- and post-settlement processes. *Oceanography and Marine Biology: An Annual Review*, **32**, 65–109.

Ormond, R.F.G., Gage, J.D., and Angel, M.V. (ed.) (1997). *Marine biodiversity. Patterns and processes.* Cambridge University Press, Cambridge.

Page, H.M. (1997). Importance of vascular plant and algal production to macro-invertebrate consumers in a Southern California saltmarsh. *Estuarine and Coastal Shelf Science*, **45**, 823–834.

Paterson, A.W. and Whitfield, A.K. (1997). A stable carbon isotope study of the food-web in a freshwater-deprived South African estuary, with particular emphasis on the ichthyofauna. *Estuarine and Coastal Shelf Science*, **45**, 705–715.

Paterson, D.M. (1986). The migratory behaviour of diatom assemblages in a laboratory tidal micro-ecosystem examined by low temperature scanning electron microscopy. *Diatom Research*, **1**, 227–239.

Paterson, D.M. (1989). Short-term changes in erodibility of intertidal cohesive sediments related to the migratory behaviour of epipelic diatoms. *Limnology and Oceanography*, **34**, 223–234.

Paterson, D.M. (2000). The fine-structure and properties of the sediment surface. In *The benthic boundary layer* (ed. B.B. Jørgensen and B.P. Boudreau). Oxford University Press, Oxford.

Pearson, T.H. and Rosenberg, R. (1978). Macrobenthic succession in relation to organic enrichment and pollution of the marine environment. *Oceanography and Marine Biology: An Annual Review*, **16**, 229–311.

Pethick, J. (1984). *An introduction to coastal geomorphology*. Arnold, London.

Pimm, S.L., Lawton, J.H., and Cohen, J.E. (1991). Food web patterns and their consequences. *Nature (London)*, **350**, 669–674.

Pinn, E.H. and Ansell, A.D. (1993). The effect of particle size on the burying ability of the brown shrimp *Crangon crangon*. *Journal of the Marine Biological Association of the UK*, **73**, 365–377.

Pollock, L.W. (1998). *A practical guide to the marine animals of northeastern North America*. Rutgers University Press, New Brunswick.

Pomeroy, L.R. and Wiegert, R.G. (ed.) (1981). *The ecology of a salt marsh*. Springer-Verlag, New York.

Potter, I.C., Beckley, L.E., Whitfield, A.K., and Lenanton, R.C.J. (1990). Comparisons between the roles played by estuaries in the life cycles of fishes in temperate western Australia and southern Africa. *Environmental Biology of Fishes*, **28**, 143–178.

Potter, I.C., Claridge, P.N., Hyndes, G.A., and Clarke, K.R. (1997). Seasonal, annual and regional variations in ichthyofaunal composition in the inner Severn estuary and inner Bristol Channel. *Journal of the Marine Biological Association of the UK*, **77**, 507–525.

Quammen, M.L. (1984). Predation by shorebirds, fish, and crabs on invertebrates in intertidal mudflats: an experimental test. *Ecology*, **65**, 529–537.

Radford, P.J. (1994). Pre- and post-barrage scenarios of the relative productivity of benthic and pelagic subsystems of the Bristol Channel and Severn estuary. *Biological Journal of the Linnean Society*, **51**, 5–16.

Raffaelli, D. and Hawkins, S. (1996). *Intertidal ecology*. Chapman and Hall, London.

Raffaelli, D. and Milne, H. (1987). An experimental investigation of the effects of shorebird and flatfish predation on estuarine invertebrates. *Estuarine and Coastal Shelf Science*, **24**, 1–13.

Raffaelli, D.G., Raven, J.A., and Poole, L.J. (1998). Ecological impact of green macroalgal blooms. *Oceanography and Marine Biology: An Annual Review*, **36**, 97–125.

Ranwell, D.S. (1972). *Ecology of salt marshes and sand dunes*. Chapman and Hall, London.

Reid, D.G. (1985). Habitat and zonation patterns of *Littoraria* species (Gastropoda: Littorinidae) in Indo-Pacific mangrove forests. *Biological Journal of the Linnean Society*, **26**, 39–68.

Reid, D.G. (1986). *The littorinid molluscs of mangrove forests in the Indo-Pacific region. The genus* Littoraria. British Museum (Natural History), London.

Reid, D.G. and Naylor, E. (1985). Free-running, endogenous semilunar rhythmicity in a marine isopod crustacean. *Journal of the Marine Biological Association of the UK*, **65**, 85–91.

Reid, R.G.B. (1998). Subclass Protobranchia. In *Mollusca: the southern synthesis* (ed. P.L. Beesley, G.J.B. Ross, and A. Wells), pp. 235–247. CSIRO Publishing, Melbourne.

Reise, K. (1985). *Tidal flat ecology. An experimental approach to species interactions.* Springer-Verlag, Berlin.

Reise, K. and Ax, P. (1979). A meiofaunal 'Thiobios' limited to the anaerobic sulfide system of marine sand does not exist. *Marine Biology*, **54**, 225–237.

Remane, A. and Schlieper, C. (1971). *Biology of brackish water* (2nd revised edn). E. Schweizerbart'sche Verlagsbuchhandlung, Stuttgart and John Wiley and Sons Inc., New York.

Revsbech, N.P., Jørgensen, B.B., Blackburn, T.H., and Cohen, Y. (1983). Microelectrode studies of the photosynthesis and O_2, H_2S and pH profiles of a microbial mat. *Limnology and Oceanography*, **28**, 1062–1074.

Rhoads, D.C. and Young, D.K. (1970). The influence of deposit-feeding organisms on sediment stability and community trophic structure. *Journal of Marine Research*, **28**, 150–178.

Ridd, P.V. (1996). Flow through animal burrows in mangrove creeks. *Estuarine and Coastal Shelf Science*, **43**, 617–625.

Riisgård and Larsen, P.S. (1995). Filter-feeding in marine macroinvertebrates: pump characteristics, modelling and energy cost. *Biological Reviews*, **70**, 67–106.

Robertson, A.I. (1986). Leaf-burying crabs: their influence on energy flow and export from mixed mangrove forests (*Rhizophora* spp.) in northeastern Australia. *Journal of Experimental Marine Biology and Ecology*, **102**, 237–248.

Robertson, A.I. and Duke, N.C. (1990). Recruitment, growth and residence time of fishes in a tropical Australian mangrove system. *Estuarine and Coastal Shelf Science*, **31**, 723–743.

Roegner, G.C. (1998). Hydrodynamic control of the supply of suspended chlorophyll *a* to infaunal estuarine bivalves. *Estuarine and Coastal Shelf Science*, **47**, 369–384.

Round, F.E. (1981). *The ecology of algae.* Cambridge University Press, Cambridge.

Round, F.E. and Palmer, J.D. (1966). Persistent, vertical-migration rhythms in benthic microflora II. Field and laboratory studies on diatoms from the banks of the River Avon. *Journal of the Marine Biological Association of the UK*, **46**, 191–214.

Rowbotham, F. (1964). *The Severn bore.* David and Charles, Dawlish.

Rowcliffe, J.M., Watkinson, A.R., and Sutherland, W.J. (1998). Aggregative responses of brent geese on salt marsh and their impact on plant community dynamics. *Oecologia*, **114**, 417–426.

Rozema, J., Gude, H., and Pollak, G. (1981). An eco-physiological study of the salt secretion of four halophytes. *New Phytologist*, **89**, 201–217.

Ruiz, G.M., Hines, A.H., and Posey, M.H. (1993). Shallow water as a refuge habitat for fish and crustaceans in non-vegetated estuaries: an example from Chesapeake Bay. *Marine Ecology Progress Series*, **99**, 1–16.

Ruiz, G.M., Carlton, J.T., Grosholz, E.D., and Hines, A.H. (1997). Global invasions of marine and estuarine habitats by non-indigenous species: mechanisms, extent, and consequences. *American Zoologist*, **37**, 621–632.

Ruppert, E.E. and Barnes, R.D. (1994). *Invertebrate zoology* (6th edn). Saunders College Publishing, London.

Russell-Hunter, W.D., Apley, M.L., and Hunter, R.D. (1972). Early life-history of *Melampus* and the significance of semilunar synchrony. *Biological Bulletin of the Marine Biological Laboratory, Woods Hole*, **143**, 623–656.

Sanders, H.L. (1968). Marine benthic diversity: a comparative study. *American Naturalist*, **102**, 243–282.

Sarda, R., Foreman, K., Werme, C.E., and Valiela, I. (1998). The impact of epifaunal predation on the structure of macroinfaunal invertebrate communities of tidal salt-marsh creeks. *Estuarine and Coastal Shelf Science*, **46**, 657–669.

Scholten, H., Klepper, O., Nienhuis, P.H., and Knoester, M. (1990). Oosterschelde estuary (S.W. Netherlands): a self-sustaining ecosystem? *Hydrobiologia*, **195**, 201–215.

Segerstråle, S.G. (1962). Investigations on Baltic populations of the bivalve *Macoma baltica* (L.). Part II. What are the reasons for the periodic failure of recruitment and the scarcity of *Macoma* in the deeper waters of the inner Baltic? *Societas Scientiarum Fennica. Commentationes Biologicae*, **24**, 1–26.

Seymour, M.K. (1971). Burrowing behaviour in the European lugworm *Arenicola marina* (Polychaeta; Arenicolidae). *Journal of Zoology (London)*, **164**, 93–113.

Smith, T.J., Boto, K.G., Frusher, S.D., and Giddins, R.L. (1991). Keystone species and mangrove forest dynamics: the influence of burrowing by crabs on soil nutrient status and forest productivity. *Estuarine and Coastal Shelf Science*, **33**, 419–432.

Snedaker, S.C. and Snedaker, J.G. (ed.) (1984). *The mangrove ecosystem: research methods.* UNESCO, Paris.

Snelgrove, P.V.R. and Butman, C.A. (1994). Animal-sediment relationships revisited: cause versus effect. *Oceanography and Marine Biology: An Annual Review*, **32**, 111–177.

Soetaert, K. and Herman, P.M.J. (1995). Carbon flows in the Westerschelde estuary (the Netherlands) evaluated by means of an ecosystem model (MOSES). *Hydrobiologia*, **311**, 247–266.

Soetaert, K., Vincx, M., Wittoeck, J., Tulkens, M., and van Gansbeke, D. (1994). Spatial patterns of Westerschelde meiobenthos. *Estuarine and Coastal Shelf Science*, **39**, 367–388.

Sotheran, I.S., Foster-Smith, R.L., and Davies, J. (1997). Mapping of marine benthic habitats using image processing techniques within a raster-based geographic information system. *Estuarine and Coastal Shelf Science*, **44** (Suppl. A), 25–31.

Stafford-Deitsch, J. (1996). *Mangrove. The forgotten habitat.* Immel Publishing Ltd., London.

Steers, J.A. (ed.) (1960). *Scolt Head Island.* W. Heffer and Sons, Cambridge.

Sullivan, M.J. and Moncreiff, C.A. (1990). Edaphic algae are an important component of salt marsh food-webs: evidence from multiple stable isotope analyses. *Marine Ecology Progress Series*, **62**, 149–159.

Summerhayes, C.P. and Thorpe, S.A. (ed.) (1996). *Oceanography. An illustrated guide.* Manson Publishing Ltd, Southampton.

Sutherland, T.F., Grant, J., and Amos, C.L. (1998). The effect of carbohydrate production by the diatom *Nitzschia curvilineata* on the erodibility of sediment. *Limnology and Oceanography*, **43**, 65–72.

Sverdrup, H.U., Johnson, M.W., and Fleming, R.H. (1942). *The oceans. Their physics, chemistry, and general biology.* Prentice-Hall, Englewood Cliffs, N.J.

Talbot, M.M.B., Bate, G.C., and Campbell, E.E. (1990). A review of the ecology of surf diatoms, with special reference to *Anaulus australis*. *Oceanography and Marine Biology: An Annual Review*, **28**, 155–175.

Teal, J.M. (1962). Energy flow in the salt marsh ecosystem of Georgia. *Ecology*, **43**, 614–624.

Teal, J. and Teal, M. (1969). *Life and death of the salt marsh.* Little, Brown and Co., Boston.

Thiébaut, E., Cabioch, L., Dauvin, J.-C., Retière, C., and Gentil, F. (1997). Spatio-temporal persistence of the *Abra alba—Pectinaria koreni* muddy-fine sand community of the eastern Bay of Seine. *Journal of the Marine Biological Association of the UK*, **77**, 1165–1185.

Thorson, G. (1950). Reproductive and larval ecology of marine bottom invertebrates. *Biological Reviews*, **25**, 1–45.

Thorson, G. (1957). Bottom communities (sublittoral or shallow shelf). In *Treatise on marine ecology and palaeoecology*, Vol. I (ed. J.W. Hedgpeth). *Memoirs of the Geological Society of America*, **67**, 461–534.

Thrush, S.F. (1986). The sublittoral macrobenthic community structure of an Irish sea-lough: effect of decomposing accumulations of seaweed. *Journal of Experimental Marine Biology and Ecology*, **96**, 199–212.

Thrush, S.F. (1988). The comparison of macrobenthic recolonization patterns near and away from crab burrows on a sublittoral sand flat. *Journal of Marine Research*, **46**, 669–681.

Thrush, S.F. and Townsend, C.R. (1986). The sublittoral macrobenthic community composition of Lough Hyne, Ireland. *Estuarine and Coastal Shelf Science*, **23**, 551–574.

Thrush, S.F., Schneider, D.C., Legendre, P., Whitlach, R.B., Dayton, P.K., Hewitt, J.E., *et al.* (1997). Scaling-up from experiments to complex ecological systems: where to next? *Journal of Experimental Marine Biology and Ecology*, **216**, 243–254.

Trenhaile, A.S. (1997). *Coastal dynamics and landforms*. Clarendon Press, Oxford.

Turner, S.J., Thrush, S.F., Pridmore, R.D., Hewitt, J.E., Cummings, V.J., and Maskery, M. (1995). Are soft-sediment communities stable? An example from a windy harbour. *Marine Ecology Progress Series*, **120**, 219–230.

Ugolini, A. (1996). Jumping and sun compass in sandhoppers: an antipredator interpretation. *Ethology, Ecology and Evolution*, **8**, 97–106.

Underwood, A.J. and Chapman, M.G. (ed.) (1995). *Coastal marine ecology of temperate Australia*. University of New South Wales Press Ltd., Sydney.

Underwood, G.J.C. and Paterson, D.M. (1993). Seasonal changes in diatom biomass, sediment stability and biogenic stabilization in the Severn estuary. *Journal of the Marine Biological Association of the UK*, **73**, 871–887.

Urrutia, M.B., Iglesias, J.I.P., Navarro, E., and Prou, J. (1996). Feeding and absorption in *Cerastoderma edule* under environmental conditions in the Bay of Marennes-Oleron (western France). *Journal of the Marine Biological Association of the UK*, **76**, 431–450.

Vance, D.J., Haywood, M.D.E., and Staples, D.J. (1990). Use of a mangrove estuary as a nursery area by postlarval and juvenile banana prawns, *Penaeus merguiensis* de Man, in northern Australia. *Estuarine and Coastal Shelf Science*, **31**, 689–701.

Wafar, S., Untawale, A.G., and Wafar, M. (1997). Litter fall and energy flux in a mangrove ecosystem. *Estuarine and Coastal Shelf Science*, **44**, 111–124.

Warner, G.F. (1969). The occurrence and distribution of crabs in a Jamaican mangrove swamp. *Journal of Animal Ecology*, **38**, 379–389.

Warren, J.H. and Underwood, A.J. (1986). Effects of burrowing crabs on the topography of mangrove swamps in New South Wales. *Journal of Experimental Marine Biology and Ecology*, **102**, 223–235.

Warwick, R.M. (1986). A new method for detecting pollution effects on marine macrobenthic communities. *Marine Biology*, **92**, 557–562.

Warwick, R.M. and Clarke, K.R. (1994). Relearning the ABC: taxonomic changes and abundance/biomass relationships in disturbed benthic communities. *Marine Biology*, **118**, 739–744.

Warwick, R.M. and Davies, J.R. (1977). The distribution of sublittoral macrofauna communities in the Bristol Channel in relation to the substrate. *Estuarine and Coastal Marine Science*, **5**, 267–288.

Warwick, R.M. and Uncles, R.J. (1980). Distribution of benthic macrofaunal associations in the Bristol Channel in relation to tidal stress. *Marine Ecology Progress Series*, **3**, 97–103.

Warwick, R.M., Clarke, K.R., and Gee, J.M. (1990). The effect of disturbance by the soldier crabs *Mictyris platycheles* H. Milne Edwards on meiobenthic community structure. *Journal of Experimental Marine Biology and Ecology*, **135**, 19–33.

Watling, L. (1991). The sedimentary milieu and its consequences for resident organisms. *American Zoologist*, **31**, 789–796.

Watt, D.A. (1998). Estuaries of contrasting trophic status in KwaZulu-Natal, South Africa. *Estuarine and Coastal Shelf Science*, **47**, 209–216.

Weslawski, J.M., Wiktor, J., Zajaczkowski, M., Futsaeter, G., and Moe, K.A. (1997). Vulnerability assessment of Svalbard intertidal zone for oil spills. *Estuarine and Coastal Shelf Science*, **44** (Suppl. A), 33–41.

Wheeler, A. (1979). *The tidal Thames*. Routledge and Kegan Paul, London.

Wildish, D. and Kristmanson, D. (1997). *Benthic suspension feeders and flow.* Cambridge University Press, Cambridge.

Williams, B.G., Naylor, E., and Chatterton, T.D. (1985). The activity patterns of New Zealand mud crabs under field and laboratory conditions. *Journal of Experimental Marine Biology and Ecology*, **89**, 269–282.

Wilson, W.H. (1991). Competition and predation in marine soft-sediment communities. *Annual Review of Ecology and Systematics*, **21**, 221–241.

Wolanski, E. (1986). An evaporation-driven salinity maximum zone in Australian tropical estuaries. *Estuarine and Coastal Shelf Science*, **22**, 415–424.

Wolanski, E., Huan, N.N., Dao, LeT., Nhan, N.H., and Thuy, N.N. (1996). Fine-sediment dynamics in the Mekong river estuary, Vietnam. *Estuarine and Coastal Shelf Science*, **43**, 565–582.

Woodin, S.A. (1974). Polychaete abundance patterns in a marine soft-sediment environment: the importance of biological interactions. *Ecological Monographs*, **44**, 171–187.

Woodwell, G.M., Whitney, D.E., Hall, C.A.S., and Houghton, R.A. (1977). The Flax Pond ecosystem study: exchanges of carbon in water between a salt marsh and Long Island Sound. *Limnology and Oceanography*, **22**, 833–838.

Wyatt, T.D. (1986). How a subsocial intertidal beetle, *Bledius spectabilis*, prevents flooding and anoxia in its burrow. *Behavioural Ecology and Sociobiology*, **19**, 323–331.

Yallop, M.L., deWinder, B., Paterson, D.M., and Stal, L.J. (1994). Comparative structure, primary production and biogenic stabilization of cohesive and non-cohesive marine sediments inhabited by microphytobenthos. *Estuarine and Coastal Shelf Science*, **39**, 565–582.

Young, B.M. and Harvey, L.E. (1996). A spatial analysis of the relationship between mangrove (*Avicennia marina* var. *australasica*) physiognomy and sediment accretion in the Hauraki Plains, New Zealand. *Estuarine and Coastal Shelf Science*, **42**, 231–246.

Glossary

Amensalism the negative effects of disturbance by one species or set of species upon another. Hence 'trophic group amensalism' is the suggested negative effect of deposit feeders upon suspension feeders.

Bed shear stress the tangential force 'felt' by the bottom when water flows over it.

Cohesive sediments those with a high proportion of small particles (<63 μm), in which the particles bind together and do not act as individual grains.

Commensalism a relationship in which one species benefits from another.

CZM coastal zone management.

$\delta^{13}C$ the ratio of the ^{13}C isotope to that of ^{12}C, used to trace the origin of organic carbon.

Dissipative beaches long flat beaches on which wave energy is dissipated in a surf zone.

DOC dissolved organic carbon.

DOM dissolved organic matter.

Endogenous rhythm rhythm that is intrinsic, triggered by pacemakers within the organism.

EPS extracellular polymeric substances found in sediments, usually mucilage from bacteria and diatoms.

Euryhaline applied to organisms tolerant of a wide range of salinity.

Eutrophication a state in which nutrient levels rise, stimulating growth of algal blooms. Subsequent decay of these may lead to de-oxygenation.

Exposure used in the sense of exposure to wave action. Thus exposed beaches receive fierce wave action because they are exposed to waves that build up over long uninterrupted areas of ocean.

Halocline a vertical gradient of salinity.

Keystone species species that are thought to have an important function in the community. The community changes drastically if such species are removed.

Macrobenthos benthic organisms greater than 500 μm in length.

Macrotidal estuaries those with a tidal range greater than 4 m.

Meiobenthos benthic organisms in the size range 100–500 μm.

Mesotidal estuaries those with a tidal range of 2–4 m.

Microbenthos benthic organisms less than 100 μm in length.

Microbial loop the route by which dissolved organic matter is utilized by bacteria and other microbes and returns to higher levels of food webs.

Microtidal estuaries those with a tidal range of less than 2 m.

Mixohaline water with a salinity between 0.5 and 30.

MPA marine protected area.

MTL mean tide level.

Negative estuary one in which water lost by evaporation exceeds the fresh water inflow, so there is a residual flow upstream.

Oligohaline water with a salinity between 0.5 and 5.

Partial predation the action of consuming part of the prey but not killing it. This part is then often regenerated, as in the siphons of bivalves.

Phi (ϕ) scale a scale on which particle size is expressed as the negative log to the base 2 of particle diameter in millimetres.

POC particulate organic carbon.

POM particulate organic matter.

Redox potential a measurement reflecting the balance between oxidation and reduction processes. It becomes negative when the sediment is very anoxic.

RPD redox potential discontinuity—a layer in the sediment where redox potential decreases rapidly, and below which the sediment is usually anoxic.

Reflective beaches steep beaches on which wave energy is largely reflected and not dissipated in a surf zone.

Salinity the dissolved inorganic matter in a particular mass of water, formerly expressed as ‰, but now as a number, e.g. 'salinity is 35.0'.

Salt-wedge estuary one in which water is highly stratified, with fresh water floating on a saline 'wedge' penetrating upstream underneath it.

Stenohaline applied to organisms intolerant of low salinities, especially those below about 30.

Supply-side ecology the processes that lead to the supply of larvae to the sea floor.

Appendix: A brief classification of selected organisms

CYANOBACTERIA (blue–green algae)	*Lyngbya*, *Oscillatoria*
BACILLARIOPHYTA (diatoms)	*Navicula* *Nitzschia* *Skeletonema*
ALGAE	
Phaeophyta (brown algae)	*Fucus* *Ascophyllum*
Chlorophyta (green algae)	*Ulva, Enteromorpha*
Rhodophyta (red algae)	*Porphyra*
ANGIOSPERMS	
Monocotyledons	*Spartina, Puccinellia, Potamogeton, Zostera*
Dicotyledons	
'mangroves'	*Avicennia, Rhizophora*
non-woody halophytes	*Aster, Salicornia*
COELENTERATA	
Anthozoa (sea anemones)	*Edwardsia*
PLATYHELMINTHES	*Convoluta*
ANNELIDA	
Oligochaeta	*Nais, Tubifex*

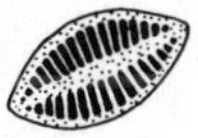

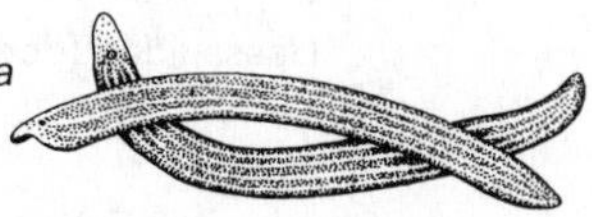

ANNELIDA (*cont.*)	
Polychaeta	
'Sedentary' forms:	
Maldanidae (bamboo worms)	*Clymenella, Praxillella*
Arenicolidae (lugworms)	*Abarenicola, Arenicola*
Sabellariidae	*Sabellaria*
Capitellidae (thread worms)	*Capitella*
Orbiniidae	*Scoloplos*
Spionidae	*Pseudopolydora*
Terebellidae	*Lanice*
'Errant' forms:	
Nereidae	*Nereis*
Nephtyidae	*Nephtys*
MOLLUSCA	
Gastropoda	
Hydrobiidae (mud snails)	*Hydrobia*
Littorinidae (periwinkles)	*Littoraria*
Cerithiidae (ceriths)	*Cerithidea*
Nassariidae (dog whelks)	*Ilyanassa, Bullia*
Amphibolidae (pulmonate mud snails)	*Amphibola*

Melampodidae (marsh pulmonates)	*Melampus*
Bivalvia	
Cardiidae (cockles)	*Cerastoderma*
Donacidae (surf clams)	*Donax*
Myidae (soft clams)	*Mya*
Mytilidae (mussels)	*Mytilus*
Nuculidae (nut clams)	*Nucula*
Ostreidae (oysters)	*Crassostrea*
Solemyidae (swimming clams)	*Solemya*
Solenidae (razors)	*Solen*
Tellinidae (tellins)	*Tellina, Macoma, Abra*
Veneridae (hard clams)	*Venus, Mercenaria*
CRUSTACEA	
Copepoda	*Calanus, Eurytemora*
Isopoda	*Tylos, Eurydice, Sphaeroma, Idotea*
Amphipoda	*Pontoporeia, Gammarus,*
	Talitrus, Corophium

CRUSTACEA (*cont.*)		
Decapoda		
Caridea (shrimps)	*Crangon, Nephrops, Callianassa, Penaeus*	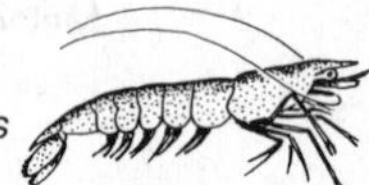
Anomura (hermits etc.)	*Emerita*	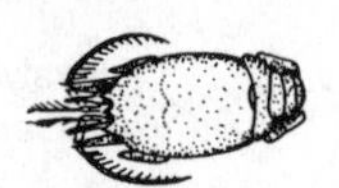
Brachyura (crabs)	*Uca, Mictyris, Aratus, Sesarma, Carcinus, Cancer, Ocypode,*	
	Callinectes, Helice	
INSECTA	*Tabanus, Bledius*	
ECHINODERMATA	*Echinocardium* *Ophiura*	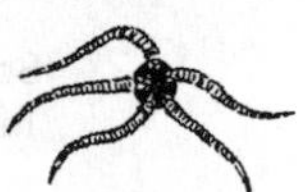
VERTEBRATA		
Teleostei (bony fish)		
Anguillidae (eels)	*Anguilla*	
Gobiidae (gobies)	*Pomatoschistus* *Periophthalmus*	
Mugilidae (mullets)	*Liza, Mugil*	
Salmonidae (salmon etc.)	*Salmo*	
Pleuronectidae (flatfish)	*Pleuronectes*	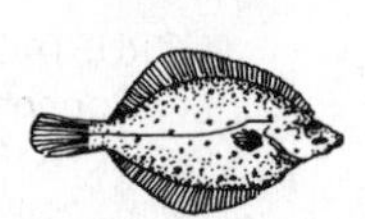
Aves (birds)		
Haematopodidae (oystercatchers)	*Haematopus*	
Charadriidae (waders)	*Tringa, Calidris*	
Anatidae (ducks and geese)	*Anas*	

Index